A POLITICAL GEOGRAPHY OF LATIN AMERICA

Latin America is currently undergoing rapid transformation. This transformation is related to changes in domestic politics and the influence of globalising socio-economic forces which have affected all Latin Americans.

This text looks at these recent changes in Latin America's power and space relations at several scales of analysis, from the global and international to the nation-state and the community. Without focusing solely on the nation-state, which in the past has been considered the pivot of political geography analysis, the political geographies of smaller scales such as localities and individuals, and broader scales such as globalising economic entities, can be given wider consideration. This pursuit of a more 'democratic' political geography in terms of weight of analysis provides a greater recognition of the ethnic, gendered, sexual and communal political geographies of the continent. It is a conscious shift of emphasis away from the military geopolitical tradition of Latin American political geography that has masked other political geographies and co-opted the sub-discipline as a tool for control and subjugation.

A Political Geography of Latin America provides a counterbalance to more traditional approaches within political geography. Students will find examples to support the 'democratic' political geography that is put forward and will also find that the text is balanced more in favour of peoples of Latin America rather than institutions, of citizens rather than nation-states, and of working hands rather than military arms. Traditional political geography is questioned and a critical democratic political geography is proposed. This latter approach to political geography should be cognisant of the widest range of power and space relations that operate within the global arena. It is a more inclusive rather than exclusive approach to the sub-discipline, recognising that *all* peoples, regardless of ethnicity and gender, and *all* places, regardless of levels of development, are integral to an understanding of global power and space relations into the twenty-first century.

Jonathan R. Barton is Senior Research Associate in the School of Development Studies, University of East Anglia.

A POLITICAL GEOGRAPHY OF LATIN AMERICA

Jonathan R. Barton

Routledge
Taylor & Francis Group

LONDON AND NEW YORK

First published 1997
by Routledge
2 Park Square, Milton Park, Abingdon,
Oxon, OX14 4RN

Simultaneously published in the USA and Canada
by Routledge
605 Third Avenue, New York, NY 10017

*Routledge is an imprint of the Taylor &
Francis Group, an informa business*

Typeset in Garamond by Solidus (Bristol) Limited

British Library Cataloging in Publication Data
A catalogue record for this book is available from the British Library

Library of Congress Cataloging in Publication Data
Barton, Jonathan R.
A political geography of Latin America/Jonathan R. Barton.
Includes bibliographical references and index.
1. Geopolitics-Latin America.   2. Political geography.   3. Human
geography-Latin America.   4. Latin America-Politics and
government-20th century.   5. International cooperation.   I. Title.
F1414.2.B34   1997
320.1'2'098-dc21   96-47376

ISBN: 978-0-415-121-89-7 (hbk)
ISBN: 978-0-415-12190-3 (pbk)

To the memory of

'S.A.J.' Wray

Intelligent, witty, sincere

CONTENTS

CONTENTS

ILLUSTRATIONS

PLATES

ILLUSTRATIONS

FIGURES

TABLES

ILLUSTRATIONS

BOXED CASE STUDIES

PREFACE

Latin America is a continent of diversity with a wide range of societies, politics and economies. Any discussion of it must therefore be preceded by the caveat that the nation-states and peoples of South and Central America, Cuba, Haiti and the Dominican Republic (which together form the political geographical units of Latin America) are highly heterogeneous. This political geography of Latin America seeks to reveal these differences and how they are manifested spatially, between places, regions and countries.

With such a starting point, it is possible to overcome the stereotyping and simplifications inherent in generalisations about the continent. Stereotypes of dictatorship and terrorism, 'banana republics', shanty town poverty and 'the Latin character' serve only to obscure the realities and complexities of Latin American life. From the Antarctic to the tropical jungles, the coastal communities to the highland villages, from the mega-cities to isolated rural existences, the political geographies of everyday lives, of localities, of cities, of rurality, and of the nation-state are too sophisticated to be subjected to all-encompassing, and often erroneous, generalisations. A more critical analysis of similarities and differences, and constructions and representations of peoples and places, leads to a clearer understanding of these sophisticated political geographies. For this reason, a critical human geography perspective is adopted in this text to examine and explain differences between peoples and places.

Apart from the clear differences between nation-states within the continent, a political geography of Latin America must also recognise diversity at other scales. While nation-states exhibit differences within and between urban centres, rural areas, and within and between the peoples and societies that exist at provincial, municipal and community levels, it is also important to recognise that individuals have personal political geographies, and that villages and cities have political geographies. For this reason, it must be recognised that an analysis of the Latin American nation-state is not the same as an analysis of Latin America's political geography. By contrasting peoples and places, statements may be made about the continent and its constituents (people and places) regarding history, social development and change, political ideology and economic organisation, production and orientation. These statements provide a particular perspective on Latin America's political geography, revealing political and spatial differences and complexities.

It is early days in the history of political geographies of the Third World (the 'South') due to the Anglocentric roots and history of the subject area in Europe and the United States. However, the processes of globalisation that are linking spaces and transgressing boundaries within and between the northern and southern hemispheres provide the context for an inclusionary political geography. This inclusionary, or democratic, political geography emphasises Southern power-space relationships to the same extent as those in Europe and the USA have been highlighted throughout the twentieth century. Recognising the need for such an inclusionary political geography that is democratic in terms of its coverage of hemispheres, nation-states, regions, races and ethnic groups, gender and sexuality is a movement towards a clearer understanding of the complex mix of political agents and spatial processes that shape the geographical organisation of the world.

In this vein, the development of Southern political geographies will lead to a more balanced view of global political geographies rather than an extension of the imperial political geographies of the late nineteenth and early twentieth centuries. These geographies placed the South in a context in which it was deemed to be complementary, rather than integral, to the processes of European and US political and economic expansionism. The movement that this text follows is one of a transition from the Great Power political geographies of the nineteenth and twentieth centuries to the global multipolar political geographies of the twenty-first century.

In moving towards a more democratic political geography, no spatial scale has been intentionally highlighted relative to others. The predominance of the scale of the nation-state emerges due to inter-nation-state dealings (international relations) and to intra-state and transboundary opposition to it. For this reason I have maintained the traditional political geography starting point, that the nation-state is the most significant organiser of global spaces. However, I stress that the nation-state is constituted by its civil society; its citizens are the keys to understanding processes of spatial organisation. Citizens are the reproducers of the nation-state, its social organisers, its labour force and its soldiers. They are also the wielders of power. They shape the nation-state that shapes them. Latin American history is littered with populist leaders, revolutionaries, dictators and activists, but it is the relationship of these individuals with civil society that has affected their ability to maintain influence and power.

This text is intended for students with a grounding in political geography who are interested in themes of Latin American political and economic development. It should also act as a guide to an understanding of power and space relations for Development Studies students since it covers many of the contemporary themes in Latin American politics and socio-economics. As a building-block for understanding the continent's geography in relation to these themes, students are directed to the Further Reading at the end of each chapter. Linking each block with others will establish a broader understanding of spatial influences in the political development of Latin America.

For assistance in the preparation of this material, I would like to thank the librarians at the University of Chile (Marcoleta), The Catholic University of Chile (San Joaquín) and the British Library of Political and Economic Science, the staff of the Latin America Bureau, students of political geography at Anglia Polytechnic University, the University of Newcastle upon Tyne, the London School of Economics and the University of Chile, and Klaus Dodds for some useful final comments. My special thanks go to Bruce Price, Sue Cunningham, Rory Miller, Nina Laurie and Nancy Nicholls for helping to shape my ideas on Latin American history and geography.

I would like to acknowledge permission from the following publishers for the reprinting of figures: John Wiley and Sons, Inc. (New York) for Martin Ira Glassner's *Political Geography*, 1996, second edition; Cambridge University Press (Cambridge, UK) for Leslie Bethell's (ed.) *Colonial Brazil*, 1987; Routledge (London) for Pascal Girot's (ed.) *The Americas: World Boundaries Volume 4*, 1994; and Lynne Rienner Publishers Inc. (Boulder, Col.) for Craig N. Murphy and Roger Tooze's (eds) *The New International Political Economy*, 1991.

Jonathan R. Barton
Norwich, September 1996

MEXICO

Mexico City ●

Havana

BELIZE   CUBA   HAITI

Guatemala City   Tegucigalpa   DOMINICAN REPUBLIC

GUATEMALA   ●   HONDURAS   Santo Domingo

San Salvador   ● Managua   Port-au-Prince

EL SALVADOR   NICARAGUA

San José   Panamá   Caracas

COSTA RICA   ●   VENEZUELA   GUYANA

PANAMA   SURINAME

Bogotá   FRENCH

Quito ●   COLOMBIA   GUIANA

ECUADOR

PERU

● Lima   BRAZIL

● La Paz

BOLIVIA   ● Brasilia

PARAGUAY

● Asunción

CHILE   ARGENTINA   URUGUAY

Santiago ●   Buenos Aires ●   ● Montevideo

Latin America: nation-states and capital cities

1

POWER TO THE PEOPLE?
Political geography and Latin America

The writer of a geographical work on South America or on any part of it, unless he is setting forth the results of his own personal observations, has a more difficult task than the historian who treats of what has happened there. The Spanish and the Portuguese, both in Europe and in the New World, have been more interested in describing characters and events than in studying the physical milieu and its influence upon human activities. Individual authors and learned societies in the Iberian Peninsula and in South America have had a great deal to say about what people did and said but are mainly silent about places.

(Shanahan, 1927; eleventh edition, 1963, viii)

POLITICAL GEOGRAPHY, LATIN AMERICA AND THE 'SOUTH'

The globalisation of political geography

From its stirrings as an academic sub-discipline over a hundred years ago, political geography has preoccupied itself with the organisation and activities of nation-states. The focus has been on the spatial dimensions of nation-states and the power relations that are established within and between them. Traditional definitions of political geography are wide-ranging but they generally encompass the same key characteristics: the nation-state, territory, inter- and intra-state relations. These characteristics were derived from the late nineteenth and early twentieth century when political geographers were writing principally from the perspective of their own nation-state in terms of geopolitical relationships with other states. An understanding of the historical context in which the early political geographers were writing is important in order to interpret their theories. However, the late twentieth-century political geography landscape is significantly different from that of the late nineteenth century and the first half of the twentieth century and requires a broader set of characteristics.

The origins of political geography are to be found in Europe and the United States of America in the writings of men such as Friedrich Ratzel, Rudolf Kjellén, Halford Mackinder and Alfred Mahan at the end of the

1

nineteenth century (see Parker, 1985; Glassner, 1996; Taylor, 1993a). The influence of these writers is reflected in political geography's evolution during the twentieth century. Europe and the USA have continued to be the foci of political geography research and it is writers from these areas that dominate in terms of international publishing. The reason for the persistence of this geographical bias is that the two World Wars and the ensuing Cold War turned the attentions of political geographers in the northern hemisphere (the North) in on themselves due to the proximity of nation-state conflicts and turbulent power-space relations that were ripe for geopolitical analysis. At the same time, the southern hemisphere states (the South) were being represented predominantly in terms of their relationship with Cold War superpowers.

The ending of the Cold War in the late 1980s with the Eastern European democratic transitions and the fragmentation of the Soviet Union has provided political geographers with an opportunity for a 'refocusing' of their research activities. The globalising processes of the post-Cold War period have created the need for a movement from a centring of research on Europe and the USA towards a global perspective on power and space. This refocusing process has also been prompted by profound changes in the South. Increasing political and socio-economic turbulence has challenged the centripetal forces of European wars and the Cold War in terms of academic attention within the North.

The most important of these changes has been the decolonisation process that has taken place throughout the South. This process of gaining independence from former imperial powers has led to a proliferation of countries seeking to establish economic, political and cultural autonomy and a reorganisation of their international linkages. The reorganisation is an attempt to break from their colonial pasts to construct post-colonial national political economies (Corbridge, 1993). The impacts of this decolonisation process have been far-reaching and have been felt within the North as well as the South, shaping changes in economic as well as political relations.

Before 1945, political geographers based in Europe and the USA were working within a web of imperial and colonial ties that linked up the spaces of the world.[1] This was their framework of analysis and the imperial powers were central to their perspectives on the world. The decolonisation process changed this predominant framework. Themes of globalisation, prompted by transnational actors and processes moving between nation-states, have led to a more fluid global arena of political forces (including nation-states and movements within civil society), economic forces, and social and cultural forces (such as environmental and social movements). Within this global framework, imperial ties have been weakened. However, at the same time new ties have emerged which are allied with the globalisation of finance, trade and economic activities, and the operations of supranational bodies such as the United Nations and the World Bank.

2

These new ties can be described as neocolonial links since the direct political control of formal imperialism has been replaced by economic influences and control emanating from the North. The G7 (group of seven leading economic powers) is an example of this continuing Northern predominance. While the traditional colonial relations of empires were distinctive in that they were established via inter-state political and military interventionism, neocolonial relations are based on state and non-state actors principally utilising the tools of financial and commercial interventionism. Military intervention has also been employed, as in US–Latin American relations. Within the globalised context of neocolonialism, the manipulation of armed force is being replaced gradually by the manipulation of market force. It is no longer the links between nation-states *alone* that are important, but also those between firms, banks, multilateral agencies and other independent organisations and movements.

Latin America and the South

The role of the South within this framework is critical to an understanding of the functioning of global relations. This was recognised within the discourse of the North–South divide that emanated from the reports of the Brandt Commission (1980, 1983) following its research into issues of global development. If the findings and recommendations of the Brandt Commission are to be taken seriously, political geography must be wholly inclusive of the South due to the complex interrelations between North and South, in terms of production and consumption, health and welfare, and the environment. The South is neither an 'add-on' nor 'peripheral' to the whole if one recognises these interrelationships. For this reason, power-space relationships in the South are deserving of comparison and critical analysis on numerous analytical levels, from the inter-municipal to the global. By adopting an inclusionary perspective of what constitutes the global, the South and its internal diversity can be highlighted and worked into theories of how the geopolitical economy (international political economy and its spatial dimensions) is changing.

Comparisons of power-space relations within the South will undoubtedly establish a better understanding of what constitute political geographies of the South as different from the North or as complementary to them. As one of the key regions within the South, Latin America's political geography contributes to an understanding of global political geographies, inclusive of the South, rather than excluding it as in the past.

Political geographies of the South are relatively scarce and build on the regional and geopolitical bases of the sub-discipline. The following three texts reveal the coverage of particular Southern regions. E.A. Boateng's (1978) *Political Geography of Africa* focuses on regional divisions of the continent and themes associated with administration, colonialism and identities linked with nationhood. This contrasts somewhat with Drysdale and Blake's (1985) *The Middle East and North Africa: A Political Geography*, which takes a look at issues of regional organisation and conflict between nation-states and 'critical issues' such as the Arab–Israeli conflict and petroleum resources. *Southeast Asia: Essays in Political Geography* written by Lee Yong Leng (1982) provides yet another selection of political geography themes, highlighting maritime geopolitics, regional identities and the geopolitical economy of the Association of Southeast Asian Nations (ASEAN) regional trading bloc. Beyond these texts, there is relatively little coverage of the South in political geography, especially if discussions relating solely to boundaries and frontiers are discounted. The geographical coverage of the articles in *Political Geography Quarterly* (later *Political Geography*) and other leading international geographical journals supports this observation.

As with other Southern regions, Latin America is diverse and complex in its geography. The complexity of Latin American societies and spaces is characterised within the continent's contemporary literary genre of magical realism, embodied in the works of Gabriel García Márquez, Isabel Allende and Alejo Carpentier, and in the folklores and traditions of indigenous societies.[2] This range of literature and cultural practices provides a clearer mental image of the Central and Southern Americas than any map since they are multi-dimensional, physical and metaphysical, historical and contemporary, reflecting the violently real, deeply spiritual and strikingly polarised features of Latin American societies. Latin American spaces, and thus geographies, also reflect these complexities. However, what the map (the geographer's traditional tool) offers that a study of the societies *per se* cannot, is to provide a spatial and physical context for the activities of these societies.

By locating societies in spaces and by recognising that they exist with, influence and are influenced by their environments, it is possible to establish an understanding of the socio-spatial frameworks within which individuals, communities and societies operate. More important to the *political* geographer are the relationships of power within and between these social actors (and their institutions), the processes that link people with spaces, and the changes that take place between the two.

While Latin America does not figure greatly within political geography literature, beyond the study of boundary geopolitics that has preoccupied Latin American political geographers (see Chapter 2), several key Anglo-European political geographers have undertaken research in the region. In 1907 and 1911, the North American political geographer Isaiah Bowman travelled to Latin America on expeditions under the auspices of Yale University and the American Geographical Society.[3] The research undertaken was wide-ranging, covering physical form and economic production and containing references to the cultural make-up of Andean societies. These judgemental references were typical of the cultural and anthropological

4

perspectives of the time. However, Bowman (1914, 172) did succeed in making the important link between geography and development, conceding that the difficulties of securing a livelihood under harsh conditions explained much of the 'low stage of development'.[4]

Carl Sauer, another renowned US-based geographer, conducted fieldwork in Latin America (principally Mexico) from the 1920s to the 1950s, making important links between geography, anthropology and history (West, 1979). Eighty years of geographical interest and Latin American 'development' later, this text attempts to bring together the geography of Latin America with issues of development and their power-space (political geography) dimensions.

Defining spaces

From the outset, there are two definitional obstacles to overcome. First, there is the locational reference to Latin America. The word America is commonly used within English language literature to refer to North America. There is a clear element of ethnocentricity within this linguistic usage since the Americas are continental reference points inclusive of North, Central and South.[5] The manner in which the word America has been linguistically usurped reflects the self-centring of the Anglocentric world with reference to the two American continents, reinforcing constructions of US hegemony within these regions.

Aníbal Quijano and Immanuel Wallerstein (1992, 549) note that the geosocial construction of the Americas came about as a result of South and Central America's insertion into the European-based capitalist economy. They continue that, 'There could not have been a capitalist world-economy without the Americas.' It is this construction of the Americas, as emerging as a result of its insertion into the capitalist economy, as Europe's 'New World' and the USA's 'backyard', that has led to Central and South America being constructed as peripheral to other powers. This construction of peripherality has played a critical role in the development of Latin America in terms of its political and commercial relations with nation-states in the northern hemisphere.

A further caveat to the use of America as a term must be the recognition of the territorial integrity of the continents prior to recognition and conquest by Amerigo Vespucci and Christopher Columbus.[6] The German geographer Martin Waldseemüller was the first person to coin the term America, in honour of Vespucci, in a map of the 'New World' published in 1507 (Williamson, 1992). With regard to these points, the term *Latin* America refers to the post-Iberian conquest period, the period after Columbus's 'discovery' when Latin-based languages (Spanish and Portuguese) were brought to Central and South America; the term Latin America was first coined in France during the 1860s (Braudel, 1994).

The long history of the Americas prior to the conquest by Columbus in

1492 is referred to as the pre-Columbian period of the Americas and encompasses 20,000 years of social change and political geography. This has to be compared with the 500-year timespan of the Iberian and independence periods. The use of Columbus's conquest as a reference point for the political, cultural and socio-economic changes within the Americas reveals the impact of the Iberian conquest and colonisation within the history of the continent: from the European perspective it is a historical juncture characterised by ethnocentric constructions such as 'discovery' and 'the New World'; from the Amerindian perspective, it is a historical juncture characterised by the initiation of Iberian genocide and exploitation.[7]

The second definitional issue is the terminology used to refer to the development condition or status of Latin American societies. Is Latin America: Third World? Developing World? Less Developed World? Underdeveloped World? Newly Industrialising? Low Income? Middle Income? Low Consumption? South? Clearly there is little right or wrong within the semantic minefield of terminology due to the range and complexity of socio-economic situations, and also factors of quality of life (Wolf-Phillips, 1987). I use the term South for the following reasons: it is a geographical reference to the southern hemisphere rather than a relative socio-economic reference; it is broad and therefore acts as a geographical 'catch-all', avoiding attempts to classify, rank or group countries or regions. The obvious criticism is that Australia, New Zealand and South Africa are also southern hemisphere nation-states and that many countries with similar development characteristics within the South are located in the North. All universalising statements are liable to be flawed, so the defence follows that the South is a geographical reference within which the *majority* of countries share *similar* environmental, social, cultural, political and economic development contexts and conditions.

The South definition is neither precise, inclusive nor exclusive, and as such it reflects the heterogeneity of Latin American circumstances. Rather than attempting to justify the term, it is better to point out that there are many and various contradictions and ambiguities within this universalisation. The definition should continue that a recognition of difference within the South is as important as a recognition of the difference between North and South. Difference characterises individuals, communities, provinces, nation-states and continents as much as hemispheres. An attempt to group and classify agents (such as people) and structures (such as nation-states) should not obscure the differences within and among them. Searching for uniformity should not lead to a blindness of difference. There are many, varied political geographies within the continent and none is more justified or valid than others, they are merely different.

This introductory chapter provides a context for the study of Latin America from a political geography perspective. By looking at political geography's central themes of power and space, the evolution of the sub-discipline itself and theories for the understanding of Latin America, a framework is constructed for the following chapters. Chapter 2 addresses the historical evolution of power-space relations relating to social and economic changes within Latin America and in international relations. Three historical periods are discussed: the pre-Columbian period (prior to Spanish and Portuguese conquest) to the turn of the sixteenth century; Iberian imperialism to the early nineteenth century; and the post-independence period to the present day.

Chapter 3 brings these historical themes up to date by looking at contemporary issues relating to geopolitics and geoeconomics within the continent. The fourth chapter is an analysis of the Latin American state, particularly with respect to the stability of political systems, forms of control over urban and rural areas, and issues of regional development. Chapter 5 provides an overview of the social and cultural diversity of the continent and its significant role in shaping late twentieth-century political geographies. Diversity in terms of social movements, ethnicity, nationhood, gender and sexuality are raised in this context. The concluding chapter advocates a shift in the theoretical balance within political geography as regards the South, and also new directions in studying Latin American political geographies to reflect the changing nature of power-space relations within the continent over a broad range of scales of analysis.

POWER AND SPACE

Beyond the nation-state

Traditionally within political geography, the spatial unit of analysis has focused on the nation-state (Johnston, 1989).[8] Contemporarily, analysts are turning their attention increasingly to other scales of analysis that tie in with and often transgress nation-state delineations, from global spaces that include and transcend the inter-nation-state system (the international) to home space and personal spaces. The recognition of more localised scales of analysis in terms of power and space brings people as units into clearer focus, rather than as national populations. This 'scaling-down' helps to emphasise differences within populations and leads to a more 'democratic' political geography of representation and recognition of difference. This can be contrasted with the homogenisation of peoples and spaces associated with the traditional power-unit structure of the nation-state.

What is more important than the particular scale of analysis is the fact that issues of power and space are common to all scales, from the global to the nation-state to the community. For this reason, it is the relationships between these two pivots of political geography that are significant. A definition of political geography that provides a framework to incorporate these is: power relationships between people(s) that impact upon and are affected by the organisation of space. This definition tries to incorporate the key traditional characteristics of political geography: the nation-state, territory, inter- and intra-state relations. By looking at people, power and space (and the processes that interrelate them), the nation-state, territorial issues and relations are integrated within this wider definition. The interrelationships of these elements may be conflictive or constructive, controlling or empowering. They are many and varied and operate along a spectrum of spatial scales from the personal space of the individual to the global. The difficulty of definition then comes down to what is meant by space and power.

Perspectives on space

Geographers are spatial analysts, but what is meant by the term space? Associations with spaces and emotional attachment to spaces are critically important to people (Tuan, 1977). These spaces are physical units, such as a room, a home, a neighbourhood (these spatial units are often referred to as places), or the territorial space of a nation-state or continent, Paraguay and South America for instance. Outer space is another scale which transcends many traditional geopolitical conceptions of space (Santis Arenas, 1985). They may also be more abstract, constructed spaces such as personal spaces or the cyberspaces of the electronic media such as the Internet. An important feature of spaces is that they are not fixed and that to analyse them it is necessary to delineate space for analysis at various scales.

By taking a broader view of spaces, the sites of potential conflict and appropriation, such as a frontier area or a shanty settlement, become clearer. It is in these sites that the state can be made or unmade relative to alternative non-state actors and movements (Fals Borda, 1992).[9] For example, the federal Mexican state seeks to assert its authority over the state of Chiapas, while non-state actors (insurrectionists) and movements (the insurrectionist army and popular support groups) try to usurp control within the area, effectively unmaking (destabilising) the state. These opposition actors and movements, that seek to undo the state or dismantle it in some way, are as important as the state itself. Democratic political geographies, based on 'reformist' conceptions of space that move beyond the focus on the state and its mechanisms of control, reveal the role of the important forces of state deconstruction as much as those of construction (Rumley, 1991). The revolutions in Cuba (1959) and Nicaragua (1979) are examples of the potential power of non-state movements to undo and overturn the existing state.

In the context of Latin America in the late twentieth century, a democratic political geography perspective becomes especially relevant and necessary due to the instability of states and polities. To move 'beyond' the state and its control mechanisms is to address freedoms within states, rather than the predominant focus on inter-state activities such as international boundary disputes (Gottmann, 1982). It is principally in the area of intra-state activities that a more democratic political geography and an emphasis on localisation (alongside the national and global) becomes increasingly important.

Contrasting with the micro-political geographies of intra-state units, the local, and the traditional nation-state, there is also the broader canvas to be painted at the world level. This has been termed the global picture, within which the local is itself highlighted, as in the term 'global village' which suggests the close relationship between macro- and micro-spatial scales. While political geography has moved around the spatial pivot of the nation-state, scaling up to geopolitics, the power-space interrelations of the inter-nation-state system, and down to localities, intra-state spatial analysis, there has been a rapid evolution during the 1990s of a recognition of the 'global' within the subject area and the impacts of processes active at this level, globalisation.

The use of the term globalisation in this text is a reference to the actors, institutions and processes that operate around the world, within and across nation-state boundaries. The international system is globalised in that there is a network of nation-states around the world that deal with one another diplomatically, politically and economically. However, transcending the territorial basis of this nation-state network are the activities of a large number of economic and social actors and processes. These include the global capital markets through which finance flows round the clock via telecommunicated transactions. The markets utilise advanced technology to exploit and transgress the territorial confines of nation-state restrictions, responding sensitively to political and economic news from nation-states and firms around the world in order to find the safest or highest return destinations for capital.[10] This process can be described as deterritorialisation.

Deterritorialisation suggests that flows of capital and technology across boundaries render 'national' spaces less significant in economic locational decision-making (Corbridge, 1994). While this term emphasises the internal national space, the sovereign territory, the terms 'borderless' (Ohmae, 1990) or 'transboundary' suggest that it is at the outer limits of territories, the interface between territories, that deterritorialisation bears greatest relevance. In the same vein, Richard O'Brien (1992, 99) advances the idea of a 'seamless' world. This process of deterritorialisation has given rise to the regional dimension of spatial organisation and capital accumulation whereby particular regions, such as the south-east of Brazil and the Greater Buenos Aires region of Argentina, have become centres of capital accumulation within the national territory. With the predominance of certain regions over others in terms of levels of production and consumption, nation-states reveal themselves to be highly polarised spaces in terms of development. This is an outcome of deterritorialisation in that the movement of firms and capital is less constrained by boundaries, and a clustering of firms and capital in particular areas occurs for reasons that include economies of scale and comparative advantage.

The critical role of economic activities in the globalisation process means that it has been characterised by the activities and movements of firms and finance across state territorial boundaries. While state territories are relatively static, global forces are marked by their fluidity in terms of location, production and trading patterns. Apart from, but often linked with, firms and finance are the social and cultural complexities of globalisation. Migration and communications have led to increasingly globalised connections between societies and cultures, within the so-called 'global village', comprised of complex webs of societies, cultures, economies and polities. Although this term has failed to acknowledge the full extent of cultural pluralities within the global space, it is true to say that in terms of media and information flows and the globalisation of production and consumption, the interconnectivity of global societies and cultures has increased markedly since the end of the 1970s.

Over 170 nation-states were represented in the 1996 Olympic Games in Atlanta, USA. Over 4 billion people watched some of the Games on television, many more read news about it, and others heard about it as information passed from community to community, individual to individual; the same was true of the 1994 World Cup football championship held in the USA. These are the fêtes or social events of the 'global village'. Technologies are the enabling tools for the compression of times and spaces which initiate broader, not necessarily deeper, interactions. These technologies, and obviously their users, are the means which account for the interconnectivities within and between societies and cultures, and also for their conflicts, fractures and divisions.

Geometries of power

While spaces provide the dimensions for political geographical analysis, it is power and its conceptualisation that determines relations within those dimensions. Political geography establishes the connections between power and space but often fails to engage with analyses of what power is and how it operates, and it is sometimes unclear how spaces change power relations and are changed by them. Power is most commonly analysed within the context of the nation-state and its relations with other actors, such as civil society, and other nation-states. Within a more democratic framework that goes beyond the nation-state, power operating within the home and at a global scale of analysis must also be accounted for.

Power is often broadly defined within geography in a relational context. Political geography has adopted discourses on power from political science and international relations (Nye, 1990; Hart, 1976), reflecting the multi-dimensional components of power, from behaviour, to decision-making, issues and interests, and conflict outcomes. The emphasis tends to fall on inter-state relations or intra-state relations based on institutional forces (see

Figure 1.1).[11] On a localised and personal level, Peter Taylor (1993a) employs the model established by Schattschneider (1960) who uses the example of two children in the school playground entering into conflict. He advocates

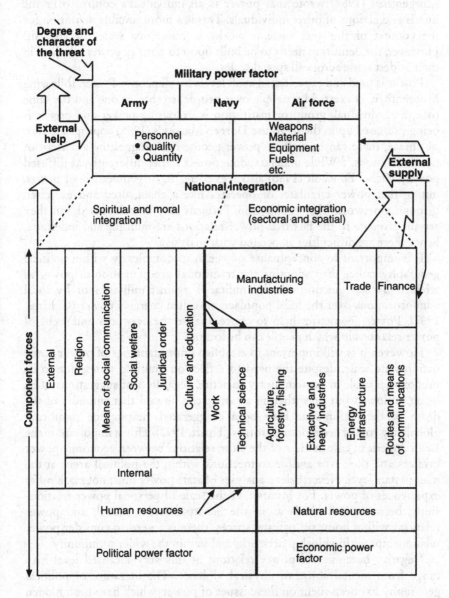

Figure 1.1 Power factors and component forces

Source: Glassner (1996). This diagram is based on a representation of power by Argentine General Juan E. Guglialmelli (translation by Glassner)

that the outcome of the conflict will depend on the support generated by both sides. This is effectively conflict resolution via an analysis of the support base. Among international relations scholars theorising on power, Hans Morgenthau (1967) wrote that power is an individual's control over the minds and actions of other individuals. This is a more suitable definition for the context of this text since it works at numerous scales of analysis. However, the definition needs to be built upon to portray power relations in their widest sense across all spatial scales.

Power is used and experienced at all levels by all people. Power, following Morgenthau, is exerted by people over people, so that, at one and the same time, an individual, group or institution is exercising power and power is being exercised upon them. To use Doreen Massey's (1995) appropriate turn of phrase, these can be termed 'power-geometries', suggesting a pattern of power relations.[12] While understanding power rather differently, as diffused power, Michel Foucault (1980) also sees patterns, or geometries, of power, noting that power circulates or operates like a chain, used and exercised through a network-like organisation. Patterns may be recognised, but they are subordinate to the fact that power relations are multiple and indefinite, hence there are difficulties associated with analysis.

It is important to conceptualise power in its complexity within political geography rather than adopting the traditional preoccupation of power as wielded by state institutions, by militaries against militaries or by local administrations over the local populace, so-called coercive power (Clastres, 1987). Power geometries help to conceptualise the networks and webs of power relations in which people can be located.

However, it is still important to establish understandings of power at the nation-state scale, despite the need to reflect on multiple scales, due to its predominant role in contemporary societies, regardless of suggestions of its being hollowed out or withering away. It can be said that rumours of the death of the nation-state have been exaggerated, despite the significant globalising role of geoeconomic forces (Thrift, 1992). This is important in the Latin American case in view of the close relations between economic power brokers and their role and/or connections within the political arena at the nation-state level. Nevertheless, analyses of state power must not mask other experiences of power. For instance, at the scale of personal power relationships, between individuals as in the Schattschneider model, and power relations within home or 'private' spaces, there is a need to consider power relationships within kinship networks and within the wider community.

Negative outcomes of power relations at this very localised level may range from mental abuse to physical violence. The literature of political geography has been silent on these issues of power, which have been hidden by the predominant scales of analysis, the nation-state and the international system. What happens in the so-called 'private' sphere, the home for example, is more 'real', recognised, understood and important for the

12

majority of individuals than international geopolitics, which they may be unaware of. This is especially significant in terms of gender relations, as highlighted in literature on Latin American women (see Radcliffe and Westwood, 1993; Küppers, 1994). The political geographies of individuals and their daily experiences of power and space relations are as valid and significant as other aggregated scales of analysis.

It is only through a wide interpretation of power geometries that spatial change can be understood in the context of Latin America. For example, the ideological power of Catholicism within Latin America is often unaccounted for within analyses of the state and military power. However, Catholicism and other evangelising Christian movements are arguably more powerful than the state in terms of popular mobilisation. How can religion be accounted for in terms of people's power-space relations? For the most part, the state seeks justification from the Church. The Latin American author-itarian states of the 1970s and 1980s sought legitimation via their links with the Church. On the other hand, the secular nature of the Cuban state provides a striking contrast within the Latin American context. What this reveals is the need to consider a wide range of agents of power in terms of the ways in which they shape the development of spaces.

Also with reference to dictatorships and other authoritarian regimes, one must take into account the power of fear. While the tortured and *desapar-ecidos* of Brazil and the Southern Cone and the massacred victims of struggles in the Andes and Central America reveal the power linked with armed struggle, another more universal power becomes evident. This is the power of fear, encompassed by Morgenthau's definition. The mental scars of Latin American political change have disfigured Latin American societies. Uncertainty, insecurity and trauma have played an important role in societal development and collective psyches. Everyone has been touched by the power of fear: those supporting certain types of regime from their fear of terrorism; those opposed to them in fear of institutional violence.

The spatial repercussions of fear and associated actions have been voluntary and enforced exile, the loss of freedoms in public spaces, the banning of protests and the imposition of curfews. These are a few examples of the physical spatial outcomes. However, the geographies of fear and trauma are mapped out just as significantly within peoples' minds. Lessie Jo Frazier and Joseph Scarpaci (1993) argue that regimes of terror were mapped onto urban spaces in the Southern Cone during the 1970s and 1980s, and that social movements, particularly human rights movements, have contested these constructed 'topographies' of fear. Similar situations have developed in rural areas of the Andes during the 1980s and 1990s in response to terrorism and state violence.

It is the recognition of alternatives to state-exercised power that leads to new opportunities for social and political, and thus spatial, changes within Latin America. As Orlando Fals Borda (1992, 311) notes with reference to social movements in the continent and their alternative forms of organisation and action:

> The Machiavellianism of force and maneuvering, the Hegelian excuse of abuses by great men, and the thesis of the means justifying the end, including power itself, are called into question.

Latin American political geographies are moving beyond the predominant post-independence emphases on the seizure of power in the traditional Marxist-Leninist revolutionary sense or that of the authoritarian *golpe de estado*. There is growing recognition that power resides, or at least should reside, in civil society with citizens themselves. By mobilising from the grassroots and empowering people via participation in opposition to the state, the state loses its legitimacy and its power. Its only recourse becomes coercive power, often in its most brutal form.

The implications of changing emphases in where sites of power can be located within the intra-state context impacts upon issues of Latin American development. Arturo Escobar (1995) refers to the discourse of development as having power in itself, employing Michel Foucault's understanding of power. He stresses the need to progress to a post-developmental era in which to re-examine the cultural contexts of poverty and reflect on the constructions of social power that lead to social injustice and inequalities.

The constructions of development and social needs have been important in establishing relations of power between those considered as having the tools of development and those receiving the fruits of development. This has been a strong component of the power of Latin American nation-states and how these nation-states engage with other states and agencies within the international and global contexts. Once development is recognised as coming from people at local levels rather than being distributed by nation-states and agencies as a commodity, the contexts of poverty, development and power are altered considerably.

THEORETICAL ADVANCES IN GEOGRAPHY AND POLITICAL GEOGRAPHY

The revival of political geography

As sub-disciplines, human geography, political geography and geopolitics are little more than broadly bounded spaces within which research takes place and literature can be located. As academic constructs, the degree to which these are constraining or liberating constructions is debatable. Within this text, human geography is understood to be a term that encompasses the research undertaken relating to individuals, groups and societies and their activities within space and how space (and perceptions of space) impacts upon these activities. Such a broad definition provides considerable leeway for the subdivided components that together constitute human geographical study. Among these, political geography has developed an increasingly important role since its academic (re)emergence during the 1970s.

While geography curricula incorporate social, cultural, economic and political geographies, the situation of geopolitics is less clear. As the forerunner of political geography, dating back to the late nineteenth century, geopolitics as an academic discipline diminished in parallel with the rise of international relations. International relations as a discrete disciplinary area emerged during the interwar period and increased rapidly in importance after the Second World War. Since geopolitics had been discredited by its perceived relationship with Nazi expansionism, it was within international relations that themes traditionally associated with geopolitics (the state, statehood, and foreign affairs) were aired and debated.

From the 1970s political geography has had a phoenix-like revival (Muir, 1975). There has been a storm in 'the moribund backwater' in which Berry (1969) noted that political geography had been stagnating, which has prompted the desire to renegotiate the academic space of geopolitics that had been controlled by international relations. This renegotiation has been especially significant since 1989, as political geography has returned to centre stage as a forum for analyses of the fluidity of the world order in terms of its spatial dimensions (Dodds, 1994). Importantly, geopolitics as a term has returned to popular use as a tool for explaining rapid developments in nation-state fragmentation and struggles for regional autonomy, such as in Eastern Europe and the Balkans.

This return to the centre-stage of a new geopolitics, critical of state power and traditional constructions and uses of geopolitical ideas, has been a significant feature of political geography from the late 1980s. Political geography now encompasses geopolitics in its traditional and newly constructed diversity, and is inclusive of multiple scales of power-space analysis. The definition of political geography cited earlier accentuates the importance of experience, of *our* political geographies, as individuals acting within politicised spaces, as much as the traditional geopolitical, international scale of analysis. The attempt here is to establish democratic political geographies where the state is not the only nexus of power, but rather one of many.

The significance of such a 'democratic' approach to political geography is that it imparts a notion of Foucauldian power operating between people(s) within and over spaces, at numerous geographical scales. While many scholars attempt to maintain geopolitics as a separate sphere of academic attention, focusing on the international, this text puts traditional state-centric notions of geopolitics to one side in order to open up the relationships between people, power and space at numerous, interactive levels of analysis, along the spectrum personal–global.

Geopolitics therefore becomes an element within a political geography approach that seeks to establish an understanding of people, power and space

15

within and between geographical scales of analysis. Within such a framework, the bounds of geopolitics are thrown open in preference to constructing a fence around the subject area in the creation of an academic *chasse gardée* (Strange, 1991). In this way, a new geo(graphical) politics or (political) geography become synonymous, encapsulating power-space relations at multiple scales of analysis. The definitions become debatable but the basic building blocks of the sub-discipline remain the same: power and space.

A century of change

Having established a broad interpretation of political geography, an attempt to reach an understanding of changes within the sub-discipline is appropriate. Unlike sociology, politics and economics, firmly rooted in their spheres of society, government and capital, human geography's spatial perspective transcends the social sciences and also interfaces with the physical sciences, particularly within environmental sciences. This flexibility is its strength, enabling it to address the rapid advances in social and cultural theory, to respond to the demands of the gender debate and sexuality, the contemporary nature of human-environment relations, and also the 'amazing technicoloured dreamcoat' of postmodernism.

These social and cultural theoretical reflections on the *fin de siècle* global political economy offer broader perspectives than those traditionally associated with geopoliticians. The state exists in partnership with markets, technology and social movements, not distinct from them. While state-centric geopolitics was most apt as a tool for the examination of the Great Power foreign policy games of the late nineteenth century, one hundred years on the frameworks of Mackinder, Kjellén and Ratzel, which privileged geographical 'givens' such as the centralised state, require considerable re-evaluation (Giblin, 1985).[13] There is a need for political geography to confront the processes and institutions operating in the highly integrated capitalist global political economy of the 1990s. Such a confrontation requires a new set of reference points.

Traditional perspectives within political geography focused on two different strands: one, the 'organic' state school founded on the writings of Kjellén and Ratzel; the other, the geostrategic school founded on the work of Mackinder and Mahan (Glassner, 1996). Both theoretical schools worked around the notions of territoriality and international relations, but in fundamentally different ways. The 'organic' state theory was rooted in the field of environmental determinism since it sought to explain society with relation to its local environment and territorial basis. The geostrategists focused on the international system, and foreign relations within that system deriving from the defensive and offensive organisation of nation-states with respect to strategic objectives.

One striking similarity of the two schools, which ran alongside one another from the late nineteenth century to the end of the Second World War, when geostrategy emerged the stronger, was their nation-state centricity and their preoccupation with armed conflicts. The subtitle of the 1903 edition of Ratzel's *Politische Geographie*, was 'of the geography of states, of circulation and of war' (Bergevin 1989, 65). Political geographers were writing from the perspective of themselves as state nationals and were developing ideas relating to their own nation's position within the international system.

The Second World War marked the end of the dominating influence of the 'organic' state school, since it had been considered, rightly or wrongly, as integral to the *lebensraum* policy substantiated by the Munich school of geopolitics directed by General Karl Haushofer (Bassin, 1987). Geostrategy emerged as the predominant theoretical school during the late 1940s. While geopolitics, revolving around the nation-state and the international system, provided the focus of pre-Second World War political geography, the later twentieth century has been marked by a diversification of the subject area from its revival in the 1970s as a reconstructed subject area distanced from its geopolitical, especially 'organic' state, history.

Political geography during the 1960s and 1970s concerned itself principally with quantitative exercises relating to electoral geography and debates relating to Marxist political economy and its spatial dimensions. The 'cultural turn' of the 1980s opened up new avenues of philosophical and methodological investigation. Debates relating to identity and the promotion of a field of critical geopolitics have been the most significant outcomes to date. From the mid-1980s, geography experienced a paradigm shift, away from the radical Marxist preoccupations of the 1970s to a focus on the applicability of social and cultural theories with respect to the organisation of spaces and the relations of people with spaces. The cultural turn shifted academics from the more politically motivated theoretical positions of the 1970s and also promoted a critique of the positivist methodologies of the preceding paradigms (Cloke, Philo and Sadler, 1991). The resulting post-positivist, post-structuralist shift in geographical thinking has demanded new strategies and directions within political geography.[14] Critical geopolitics has provided the framework for most of these new approaches.

Critical geopolitics

Critical geopolitics emerged from the late 1980s in the work of Simon Dalby (1990a, 1990b, 1991), Gerard O Tuathail (1986, 1994) and others. What is 'critical' about this type of geopolitical research is a reflection on the ways in which geopolitical issues are represented and contextualised. Rather than the traditional geopolitical research focus on geostrategy in terms of planning and outcomes, critical geopolitics turns to the construction of the ideas behind these events and how they are represented by analysing the reading, writing and constructions of geopolitical discourses. Rather than replacing the traditional geopolitical methodologies of political geographers, critical geopolitics provides a different platform for the examination of geopolitical events and political geographies. The critical nature of this theoretical development is valid for all levels of spatial analysis, not solely international affairs as has been the focus to date.

Of significance in terms of Latin American political geography is the fact that critical geopolitics has been important in making links with development geography. Writing in the second volume of Richard Peet and Nigel Thrift's *New Models in Geography*, David Slater (1989, 28) attempted to provide an insight into how development geography might respond to the particular and changing nature of peripheral societies. Concluding, he wrote that, 'critical regional analysis of peripheral societies will stagnate unless we recast our perspective, throw off the fetters of econocentrism, and begin developing a substantive political theory of the peripheral state and social conflict'.

Although he was not writing explicitly in the context of political geography, his assessment is most pertinent to an understanding of the treatment of the South, since it raises questions about the interface between development geography and political geography. Slater's (1993) reiteration of this self-same set of fetters four years on suggests that, even under a different set of international conditions, running parallel with the current geopolitical transition, analyses of the periphery have changed little. The periphery continues to be represented principally as an object of change and devoid of agency. This perspective predominates as an outcome of the traditional Eurocentric and Anglo-American construction of the subject area (Dodds, 1993b; Kofman, 1994).

The contemporary framework of critical geopolitics and the 'new geopolitics', which seeks to provide an historical-geographical component to international political economy (Ward, 1990; Agnew and Corbridge, 1989, 1995), have emerged to confront traditional perspectives. The debates emerging from these perspectives have served to examine the situation of the South within political geography discourses (Slater, 1993, 1994; O Tuathail, 1994), in terms of ethnocentricity, academic imperialism and the representation of others (Amin, 1988; Sidaway, 1992; Slater, 1989a, 1992). Greater recognition of positionality, the bias of the observer or writer, has revealed the wider scope for more democratic political geographies of the South which are more emancipatory and empowering than previous political geographies.

Another opening that has been established within the reconceptualisation of political geographies has been the exposure of the 'hidden' geographies of minority voices in terms of race, gender and sexuality. In 1990, Eleonore Kofman and Linda Peake wrote about the prospects for a 'gendered agenda' within political geography. More than establishing a new agenda or creating new space for feminist and masculinist considerations within political geography, the impetus was for a recognition of the existence of political geographies which have been left off the traditional agenda. In the same way as feminist writers have been stressing the need to address formally 'silent histories' of women, the case has been put for 'silent geographies', or rather 'silenced geographies', of women and men that are not subsumed within the dominant typeset of state-centricity (Dalby, 1993).

Within this 'gendered agenda' project is also the need to recognise diversity, between and among women and men, in terms of sexualities, gendering influences and roles. Only with a recognition of difference can political geography include diversity within its analytical frameworks and attempt to understand relations of power and space inclusively rather than exclusively. As such, gendered political geography is neither feminist geography nor cultural geography in any explicit way, but rather a recognition of gendered identities in power-space relations.

As a fundamental outcome of the organisation and control of spaces resulting from relationships of power, a recognition of the gendered aspects of political geography is a *sine qua non* for the movement towards a broader-based, a democratic political geography. Recognition of the gendered nature of space has been slow to emerge within political geography, very much due to its history of attention to issues of inter-state conflict, state power and organisation, and a patriarchal bias to its history and geography in terms of writers and subject matter (Rose, 1993).

Humberto Maturana (1992), the Chilean philosopher, is not alone when he writes that discourses of authority, control, domination and power are essentially patriarchal and result from the defence of a patriarchal construction of an absolute or transcendent truth.[15] It is this construction of knowledge that is fundamental within this debate. While this bias is not confined to political geography, it does not diminish the need to question the structures of academic organisation, to question subjectivities and research agendas, and to challenge predominant discourses within geography, the social sciences and society as a whole. Reflections on geographical scales of analysis are good starting points for this questioning.

INSIDE OUT: A LATIN AMERICAN PERSPECTIVE

Geographies of Latin America

Anglocentrism hangs like a shroud over political geography, and questions of academic imperialism bear considerable weight in any analysis of Southern political geographies. Due to the development of Northern bias in political geography, a great deal of space has been created for contemporary interpretations from the South and of the South. Anglocentrism has led to a relative lack of attention to the non-Anglo Saxon regions of the world and a high degree of parochialism in mainstream Northern geography (Potter, 1993). David Livingstone (1992) makes the point that it is geography's close association with militarism and imperialism that has led to the selective spatial applications of the geographical tradition. Associated with this point, Paul Claval (1986) (focusing on British geography especially) puts the blame at the door of a lack of foreign language skills which is certainly critical in terms of its contribution to the lack of understanding of the writings and experiences of individuals and other societies beyond the English-speaking world. Whatever the underlying factors, Anglocentricism has led to the emergence and predominance of a particular set of perspectives, writings and constructions on and about the world.[16]

Human geography texts on Latin America written in English fall into the mould of country-by-country overviews incorporating economic, social and political themes, such as Preston James's *Latin America* (1969, fourth edition, first published in 1942) and Harold Blakemore and Clifford 'Smith's (eds) *Latin America: Geographical Perspectives* (1983), or of regional divisions of the continent linking human activities with topography, climate and geology, such as in E.W. Shanahan's *South America: An Economic and Regional Geography* (1963, eleventh edition, first published in 1927).[17] In terms of political geography, they tend to concentrate on infrastructural political issues rather than pointing out the political divisions within and between societies. A wider interpretation of what constitutes Latin American political geography is presented in this text, incorporating the cultural turn in geography that has led to a greater interest in the social make-ups of political spaces and the conflicts that develop as a result of social diversity.

To make a case for what constitutes a 'Southern' political geography requires tracing the development of political geography within the South, in this case Latin America, and the manner in which political geography is written, published and distributed. A starting point is that Latin American political geography should not be synonymous with the inter-state system and the priorities of military geopolitics, as exemplified by the numerous military geopolitical institutes within the continent and their links with European, principally German, geopolitical discourses.[18] Latin American political geography should be broadened to include the needs, demands and conflicts of communities and people(s) *within* the republics rather than solely between them.

A second point is that the Latin American state is not the western liberal democratic state of Europe or the USA. Its history, evolution and structures are not those of Northern states (Quijano and Wallerstein, 1992). One has to take into account essentially Latin American criteria such as personalism, clientelism, patrimonial politics, polarised social structures, economic frameworks and orientations, and the strength of political and religious ideologies and belief and value systems. To this effect, Latin American state formations bear closer resemblance to those of Africa and South-east Asia, exhibiting the colonial features of Northern states projected southwards, yet they are intrinsically Southern in their linking of the colonial and the indigenous and local.

An analysis of the geographical discipline within the Latin American continent reveals its emphasis on traditional human geographical concerns of urban spatial organisation and environmental considerations. Explaining the paradigm shifts within the continent and the impetus for these shifts requires an understanding of the background of the discipline. There can be little debate that Latin American geography is essentially a geography of European influence and of European immigrants rather than of the indigenous peoples. The changes within the pre-colonial, colonial and independence periods within Latin America render it a region with a strong heritage of political geography. The extremes of political ideology and political policy in terms of spatial organisation provide this political geography with considerable diversity.

Latin American political geography

The Latin American field of political geography developed from the early geographies of European natualists such as von Humboldt and Steffan from the early nineteenth century. These geographies concentrated on natural regions and natural resources, focusing on an encyclopedic approach to information collation, and were very much in the mould of travel and exploration geographies of the time, closely related to European imperialism (Reboratti, 1982; Pratt, 1992).

Human geographical issues relating to social and economic themes followed during the twentieth century at various rates, swiftly in Brazil following the establishment of the influential National Institute of Geography and Statistics (IBGE) in the 1930s, gradually in the cases of Argentina, Chile and Venezuela, and more slowly in Paraguay and Bolivia, for example (Pinchas Geiger, 1970).

In Central America, the US influence in cartography and planning delayed the development of geography. It was not until 1938 that the first Central American geography department was established, at the University of Panama (Minkel, 1970). In the 1950s, when professional geography emerged more widely throughout the continent, the influence of Vidal de la Blache and the French school was important, introducing new reflections on thought and method in geography, and also on planning and international perspectives. This paradigm shift established a social awareness in Latin American geography (Santis Arenas and Gangas Geisse, 1984–85; Toledo and Zapater, 1991).[19]

An important underlying theme within geography's development in Latin America has been territoriality. The term territory rather than space is significant in as much as territorial studies, or the organisation and ordering of territory, are understood and taught as commonplace (Vega González, 1991). Territorial studies in their widest sense provide the most common connection with political geography within contemporary geographical

21

studies. These territorial studies programmes are often oriented towards urban and regional planning. For example, in Cuba after the revolution in 1959 geographers played key roles in local and regional spatial organisation within institutes of physical planning and geography (Luzón, 1987–88).

The grounding for political geography in particular resulted from the influence of key individuals such as von Humboldt and the US geographer Bowman, also the influence of European geopolitical training within military training academies. The European academic tradition of political geography has a long history in Latin America. The focus of this tradition has consistently been on geopolitics with other debates receiving relatively little attention. The Argentinian Juan Milia (1990, 68), in an article on conceptual and methodological advances in political geography in recent decades, concedes that the fundamental objective of the paper is to 'help cover the bibliographical deficit within the country, and in the Spanish-speaking world in general'. This is an important statement, revealing that Spanish language literature has emerged mainly from military sources rather than academic institutions. A reason for this is that geography has often been swallowed up by the military information machine during times of authoritarian rule.

Beyond the military academies, geopolitics arrived relatively late to academic institutions, even in countries with highly developed European links such as Chile (Santis Arenas and Gangas Geisse, 1985).[20] Latin American political geography fits well with Yves Lacoste's reflection on the role of geography. He states that geography is, above all, strategic knowledge linked with politicians and military practices (Lacoste, 1976; Méndez, 1986). Lacoste continues that geography has been dominated by, and linked with, larger states and those exercising power over them. The political geography sub-discipline has thus favoured dominant groups and interests. It is in reaction to this background that democratic forms of political geography need to be forthcoming. This is especially appropriate within the Latin American context due to the historical predominance of authoritarianism and non-representative interest groups.

A movement within political geography and geopolitical analysis, away from traditional state-centricity towards an understanding of dialectical relationships between power and space, confronts the criticism that political geography offers little more than analysis of a particular organisational form, the state (Raffestin, 1980). Milia (1990) puts forward the need to base political geography, power-space relationships, on the grounds of political economy and its social mechanisms. He also advocates that territory should not only be understood as sustaining political forces but also as a political force itself. These points link in well with discussions of localised social movements within civil society.

Geography and development

To move political geography beyond the military confines of its history in Latin America requires an engagement with issues relating to wider development. The concept of development has been central to geographical studies and other social sciences since European geographers, principally explorers, traders and 'fortune seekers', began making statements about social and cultural organisation and physical environments elsewhere in the world. Such statements were imbued with ethnocentric cultural norms and value systems that led to a cultural hierarchisation. 'Development' and 'underdevelopment' were judged relative to the European experience of organisation and accepted practices and activities. In this way, 'Other' cultures were constructed as being relatively underdeveloped, or more commonly portrayed as backward, barbaric and uncivilised.[21]

The 'civilising mission' was the term applied to the justification of imperial expansion in order to civilise ('Europeanise') these 'Other' cultures. Civilisation, it was established, was Christian, god-fearing, and based on the European social system of morality, justice and family structure, among other factors. This 'mission', alongside the perceived resource wealth of new lands, was at the heart of the conquest of the Amerindian societies within South and Central America from the late fifteenth century.

Effectively, a new America was invented by the *conquistadores* since the pre-Columbian Americas could not co-exist with Europe (O'Gorman, 1958). America had to be remodelled as the New World, a New Europe. Pre-Columbian societies and their activities were forced to adapt to the New World model and could not live outside it. In this way, the pre-Columbian Americas were transformed into an extension of Europe. As late as the twentieth century, this same style of theoretical construction of the 'Other' world has predominated. The European model of development has prevailed as the path to be followed by other cultural groups and societies, characterised by the idea of modernisation.

Modernisation involved the reproduction of structures and processes that had been exhibited in the evolution of most European societies. The principal proponents of this school of development were Walt W. Rostow and Talcott Parsons from the late 1940s (Rostow, 1960; Parsons, 1948, 1951). Their influence dominated northern hemisphere ideas and teachings relating to socio-economic change in the southern hemisphere. Due to the relationships between states, aid and development agencies, and academic institutions, these ideas proliferated within the North and South, filtering into development policies within Latin America. The modernisation process was constructed as a unilinear path of development with a simple dualism of modern and traditional structures: Europe was the model for the modern, Latin America for the traditional.

The importance of development models in shaping thought and practice in and about Latin America suggests that political geography has to make links with theories of development. Ongoing debates about paths of development, especially alternatives to predominant models, are critically significant in terms of changes in Latin American political geographies (Brown, 1996).

It was from within Latin America that opposition to this Anglo-European model of development emerged (see Kay, 1989; Hettne, 1995). Two schools of development theory evolved that offered different strategies for the development of Latin American societies: structuralism and dependency. Structuralism, a Latin American perspective on neoclassical economics, was promoted from within CEPAL (Comisión Económica para América Latina, the UN's economic commission for Latin America, in Santiago, Chile) while dependency, a more radical critique of modernisation, emerged among Latin America's radical scholars.

Both these schools of thought emanated from a Latin American political philosophy tradition developed during the early twentieth century. The most influential proponents of this tradition were the Peruvians Víctor Raúl Haya de la Torre and José Carlos Mariátegui (Kay, 1989). While they both pursued different approaches to an explanation of the Latin American condition, they established the philosophical environment for the emergence of intrinsically Latin American development thinking. This marked a sea-change for development theory, an alternative to Anglo-European traditions. From the 1940s, Latin American theories were propounded for Latin American development. The autonomous development path followed by Perón in Argentina during the late 1940s and early 1950s provides an example of the application of this type of thinking.

While these theories did not sever the link with European academic thought regarding the development process, they set in motion the pursuit of self-reliance and the confidence of intellectual autonomy. By the 1920s, following the crisis in Europe engendered by the First World War, the role of the USA in Latin America had also become a more significant issue in terms of influence within the continent. Prior to the twentieth century, US influence had been important in Central America and the northern Latin American states but the interwar period saw US expansionism throughout Latin America which would be consolidated later during the Second World War. The pursuit of a more independent development path was a reaction to emerging US interests in the continent as much as it was a reaction to the older European influences.

The structuralist theoretical school was based on a recognition that Latin American development, or lack of it, was related to the continent's commercial relations with Europe and the USA. This idea was based on the tradition of Haya de la Torre, who had criticised imperial relations and foreign domination of Latin American economies. Raúl Prebisch, the first director of CEPAL, and the other *cepalistas* were the principal structuralists.

They worked towards an explanation of Latin America's development condition relative to Northern nation-states by emphasising inequalities embedded in trade relations (Prebisch, 1965). It was considered that the low price of Latin America's traditional exports, principally agricultural products, minerals and other raw materials, relative to higher cost imports, principally manufactures, was central to any explanation. Over time, Latin American countries' terms of trade were worsening. The presumed remedy was the promotion of industrialisation to replace imported manufactures. The policies that resulted from this idea were collectively termed import substitution industrialisation (ISI) (see Baer, 1975).

While the structuralists of CEPAL attempted to work within the boundaries of the capitalist system, embarking on reforms of the system to favour Latin American economic growth patterns by improving terms of trade and gross domestic product (GDP), the dependency theorists, *dependentistas*, were more radical in their outlook. In 'The Development of Underdevelopment' (1966), Gunder Frank situated Latin America within its historical context of mercantilist and capitalist relations and focused on the exploitative, dependent relations between the metropolitan countries and their satellites. In 1969, Cardoso and Faletto's *Dependencia y Desarrollo en América Latina* followed in the same vein and theories of dependency became established.[22]

Dependency theories revolving around the themes of historically based commercial relations within the capitalist world economy became numerous during the 1970s as different directions of theoretical explanations and responses were pursued (see Chilcote, 1974; O'Brien, 1975). Eventually the dependency school waned as an influential theoretical influence due to the emergence of neoliberal approaches towards Latin American development, suggesting that development lay in the integration of Latin American economies into the international economy in preference to more autonomous strategies.

During the 1970s, the most significant strand of late twentieth-century development theory emerged. Neoliberal thinking was based on monetarist economic theory developed by von Hayek and Milton Friedman and has formed the basis for development policies oriented towards Latin America and from within. Neoliberal policies attempt to open up national economies to the international marketplace, despite the criticisms that they are merely exacerbating conditions of low income and absolute poverty (Leftwich, 1993; Green, 1995). Strategies of market deregulation, the privatisation of public sector activities, and the opening up to foreign investment have ensured that domestic economies become more subject to external economic change. However, neoliberal economists see 'opening up' as the way to attract the necessary investment capital in order to trigger economic activities directly and indirectly. This model of development, an exposure to the vagaries of the global economic system, predominates within Latin America and provides the polar opposite to the bottom-up development strategies being adopted by many social groups and movements which have been most vulnerable to the negative consequences of the neoliberal programmes.

While the structural reformists and dependency theorists argued for alternative models of Latin American development, it was the strength of international commercial ties (principally with the USA), the predominance of the modernisation framework, and the failures of the national models of ISI that have designated late twentieth-century development patterns. The outcome has been that the modernisation project of the transfer of resources, finance and technology from wealthy areas into less wealthy areas, with the aim of promoting regional identity and territorial political participation, revealed itself to favour the élites and interest groups rather than giving rise to national development across the board. Its failure to initiate resource 'trickle-down' either socially or spatially led to the call for alternative strategies for the promotion of the endogenous mobilisation of resources and capacity for innovation and regional adaptation (Stöhr, 1987). This call has emerged not from the state and its organisers but from civil society itself, the citizens.

It is these citizens' groups, popular organisations and movements that are moving against neoliberalism as a development strategy, and those actors and institutions which promote it, by protesting and establishing self-help organisations that focus on immediate local needs. They are, in effect, bottom-up strategies of development working from the grassroots at a local level in contrast to the top-down, state-centric strategies that have failed to bring about social change and the redistribution of resources to alleviate conditions of socio-economic stress.

Bottom-up theories of development are a response to the failure of the state and international institutions to allocate and distribute resources effectively from the top. What these bottom-up theories embody is the central strategy of basic needs provision. These needs are provided by communities and movements organised within civil society. In terms of the fact that these basic needs are derived from the lowest level within the hierarchy of the state (as it is most often constructed – from the Presidency to 'the People') and are filtered down through the state system, it can be said that many movements function 'beyond the state'. State resource allocation fails to enable many millions of Latin Americans to meet their basic needs, therefore they seek to help themselves, independent of the state mechanisms that they regard as corrupt, self-interested and incompetent.

Many of these basic needs projects are assisted by nongovernmental organisations (NGOs) financed by donors from within or outside the state in question. The essence of the theories of basic needs and self-help is that projects should involve participation, a small scale, and direct action within a locality. The locality is the geographical or spatial designation of a community in this respect – the social organisation is communal and the

Table 1.1 Development theories predominant within Latin America

Time scale[a]	Theory	Influential proponents	Key features
1920s–40s	*Early theories*	Haya de la Torre Mariátegui	Early structural reformism Latin American Marxism – Ethnicity
1940s–70s	*Modernisation*	Rostow – Parsons	Unilinear development – Modern/traditional dualism – Anglo-Eurocentrism
1940s–80s	*Structuralism*	Prebisch – CEPAL Furtado	Terms of trade – ISI
1960s–80s	*Dependency*	Frank – Quijano Cardoso and Faletto	Expropriation – Exploitative relations – Core-periphery
1970s–90s	*Neoliberalism*	von Hayek – Friedman	Monetary control – Free markets – Open economies
1970s–90s	*Basic needs/ Grassroots social movements*	NGOs – Civil society	Essential needs – Community – Self-help

[a]Latin American political parties and social movements reveal that all these theoretical underpinnings still exist in the continent today. The time scales indicate the period of greatest political and social impact.

spatial scale of organisation is local. In that projects are directed at the roots of society, the people, they are commonly known as 'grassroots' organisations. The basic needs that must be met vary according to the prevalent locational and socio-economic factors, but they usually focus on food security, potable water, housing, electricity, health care, and infrastructure such as roads, drainage and other sanitation requirements.

The end of the Cold War has provided a greater space within which the new development debates revolving around sustainability, empowerment, basic needs and other so-called 'alternative' development ideas have been promoted. This sea-change is most convincingly related by Gunder Frank in a review of development literature. Frank aligns himself with the 'alternative' school, breaking from the dependency school, when he concludes that social movements, such as grassroots support structures and nongovernmental organisations, are the hope for development, particularly along the lines of the preponderance of women's groups and, 'their less hierarchical (more democratic?) organisation'. With the promoters of dependency theory adopting new positions, and the modernisation, or developmentalist, school undermined, alternative theories are very much to the fore in the 1990s (Frank, 1992).

These alternative development approaches signal a shift away from redistribution with growth and neoliberal policies of market adjustments which work on a national basis. The emphasis of basic needs and grassroots

27

is on the local, a development from below or within rather than from above (Stöhr, 1981; Taylor, 1992). Fraser Taylor (1992, 257) stresses the change in orientation of the successive development paradigms that have shaped development thinking, noting that, 'development from within is to allow local people to become the subject, not the object, of development strategies'.

What is to be gleaned for political geography from its engagement with development thinking is that intra-state dynamics define the state and that Latin American states are complex and are agents of change, as are groups of people within those states. Ultimately the issue is that of the political geography of difference and complexity, of the fact that power and space relations are not only inter-state but are very much determined by intra-state activities, by citizens, communities and societies. Taking this approach, an emphasis on the role of women and other 'hidden' social groups in these above-mentioned development strategies should assist in the awakening of political geography to its exclusionary past and the fundamental weakness of its limited examination of gender relations.

The political geographer must take heed of the development debates, especially in light of the fact that they point to the transcendence of the centrality of the state in response to the failure of the state as 'provider'. This situation raises the issues of community, localised power relations and self-help (Dalby, 1993; Corbridge, 1993). As the state becomes an increasingly problematic unit on which to base analysis, especially in the context of the South, where the state is most often self-serving and unrepresentative, the need for political geography to engage with these ideas becomes increasingly insistent. A more flexible approach to the role of economic structures and activities, both local and global, that operate within, through, and beyond the state, is necessary.

It is only through an engagement with the theories of development that have dictated patterns of socio-economic change during the twentieth century that power-space relations within Latin America can be established. The historical context of the social structures and relations within states and between states provides the framework for the understanding of the contemporary organisation of Latin American territorial space and associated power relations.

FURTHER READING

Introductions to political geography are provided by Joe Painter's *Politics, Geography and 'Political Geography': A Critical Perspective* (1995), John Rennie Short's *Introduction to Political Geography* (1993) and Peter J. Taylor's *Political Geography: World-economy, Nation-state and Locality* (1993a). Martin Ira Glassner's *Political Geography* (1996) is an encyclopaedic overview of the sub-discipline of political geography. David Preston's edited text on *Latin American Development: A Geographical Perspective* (1996) and Arthur Morris's *South America: A Changing Continent* (1995) are the most recent geographies of Latin America written in English. For approaches to Latin American development theory, see Cristóbal Kay's *Latin American Theories of Development and Underdevelopment* (1989) and Bjorn Hettne's *Development Theory and the Three Worlds* (1995). Simon Collier *et al.* have collated chapters on Latin American geography, history, economics, society and culture in *The Cambridge Encyclopaedia of Latin America and the Caribbean* (1992). Phil Gunson, Andrew Thompson and Greg Chamberlain offer detailed summaries of people and events in *The Dictionary of Contemporary Politics of South America* (1989) and *The Dictionary of Contemporary Politics of Central America* (1990).

29

2

FROM PRE-COLUMBIAN TO POST-COLD WAR GEOPOLITICS

> From [a survey] of critical boundaries and of capitals in Latin America, it seems clear that much of the area is geopolitically immature. Many boundaries in the interior are undetermined or have recently been in process of demarcation, and boundary quarrels and wars have been rife ever since the close of the colonial era.
>
> (Whittlesey 1944, 476)

PRE-COLUMBIAN GEOPOLITICS

The original discoverers

Latin American geopolitics does not begin with the construction of nation-states after independence in the early nineteenth century (or late nineteenth century in the Brazilian case). Much of the work on geography and the organisation of space within the continent does indeed tend to begin with the 'discovery' by the Europeans and suggests that the political division of the New World is still unconcluded today; this is the view of Derwent Whittlesey for example. However, while geopolitical debates and disputes are certainly not concluded over much of the continent, the spatial division of the continent among Amerindian social groups during the pre-Columbian period has equal importance for the long-term evolution of boundary and other spatial organisation issues, such as urban and economic production locations.

The pre-Columbian indigenous groups were themselves geopolitically oriented, albeit along different socially organised lines. Hegemonic groups were very much apparent during the pre-Columbian period: the Iberian colonial period and the nineteenth-century struggles for independence were not ruptures marking distinct periods of historical experience, but rather junctures of geopolitical transition within and among hegemonic groups, external and internal to the continent. In terms of establishing power relations between groups, the inter-state conflicts after independence reflect conflicts between social groups within the continent during the previous 20,000 years. The Europeans did not 'discover' a utopian paradise of socially harmonious organisations. Among Amerindian groups, particularly those subordinated by the more highly organised theocratic states of the Incas, Aztecs and Mayas for example, the pre-Columbian period was characterised by social control and extensions of territorial-administrative powers. Edwin Williamson stresses this point (1992, 38):

> America, of course, was far from being a virgin continent ... As for the original inhabitants, who had lived there for several millennia before the arrival of Columbus and had penetrated every corner of the continent, the conflicts and problems associated with historical change had not passed them by, even though theirs was a world circumscribed by the natural boundaries of the Americas. The arrival of the Europeans was but one more phase in their historical experience; by its very nature it was a traumatic phase, but perhaps not the worst in the history of many of these native peoples.

The Americas were populated by groups not autochthonous to the continents from 40,000 years ago. Theories of derivation of American peoples are numerous but generally boil down to two principal sources of origin: first, primitive groups migrated from Asia through the Bering Strait or through the Aleutian Islands into southern areas (Newson, 1996); second, southern American groups trading with South Pacific and Australasian societies may well have stimulated increased movements of people into the Americas through southern territories. As an example of the latter argument, Carvallo Ortiz (1988) notes that the Chonos of southern Chile are possibly of Polynesian origin due to their physical traits. With these movements of people, the organisation of American space and power relations associated with it followed – political geography had reached the Americas. By the fifteenth century, these geographies were well established (Denevan, 1992).

During the thousands of years that followed the initial colonisation of the American landscape, many distinct social groups emerged within geographically distinct regions, adopting localised production processes, agricultural systems and patterns of reproduction of labour. Within the Andean societies for example, northern Andean groups did not have highly developed systems of neighbourhood cooperation in AD1000 whereas central Andean groups had relatively sophisticated institutions of cooperation (Hocquenghem, 1991). While attempting to avoid the trap of geographical determinism, it is clear that distinct social groups established different systems of production and reproduction according to their geographical circumstances. This explains the wide variations in pre-Columbian socioeconomic orders and organisation, and also the contemporary distinctions between indigenous groups. Camargo Pérez (1988–89, 53) describes the Americas that Columbus arrived upon as a 'geo-social theatre' that had developed over 40,000 years from the arrival of what he terms 'proto-Asians', based on the former theory of origin. These proto-Asians and Polynesians established cultural and economic activities that contrasted with their geographically distinct origins, having been shaped by the geography and ecology of the Americas. Diverse societies with diverse characteristics emerged (Camargo Pérez 1989, 59–60):

> The human seed of America is the seed of Asia, developed alongside regional migration and movements – statures, colours and customs – derived from development in Alaskan snows, Canadian forests, US plains, Mexican valleys, Caribbean islands, slopes of Colombia, Andean sierra, pastures of Venezuela, Brazilian river systems and the open pampa of the Southern Cone.

Throughout the Central and South Americas, pre-Columbian societies could be divided into groups according to their predominant activities: pastoral, hunting and fishing. In contemporary Chile, Carvallo Ortiz (1988) categorises indigenous groups as: northern; from the longitudinal valleys; Andean hunters; and southern fishing peoples. The groups were quite different in terms of language, activities and cultures. In the central south of the continent, semi-nomadic groups predominated, such as the Guaraní whose language still dominates within eastern Paraguay. What can be established is that a geopolitical organisation of space was effective between and among social groups prior to the Iberian conquest. This organisation of space was based principally on topographical regional divisions linked with economic activities. The same was true throughout the continent where coastal, upland, plains and forest peoples inhabited distinct regions engaging in different economic and cultural activities.

Table 2.1 Indigenous peoples prior to Iberian colonisation within Chile

---

Northern: the Chango, the Atacameño, the Diaguita. These collectivities lived on marine resources. They extended over the Andes to Salto a la Rioja.

Longitudinal valleys: the Picunche, the Mapuche (Araucanos), the Huilliche. These collectivities were principally hunters and agriculturalists. All three speak the same language. The names reflect tribal geographies: picunches – northern people, mapuches – people of the land, huilliches – southern people. Another group – the *Cuncos* – left the region to inhabit the Chilean island of Chiloé.

Andean hunters: the Pehuenche, the Puelche, the Tehuelche or Patagones, the Onas. These were upland collectivities associated with hunter-gatherer activities.

Southern fishing peoples: the Chono, the Alacalufe, the Yagane. The Chonos were possibly of Polynesian origin since they have similar physiological characteristics. They established a maritime tradition preserved by the Chilotes and inhabitants of the Chiloé archipelago. All three collectivities are known as *fueguinos*, an association with the volcanoes of the region, hence Tierra del Fuego (Land of Fire).

Source: Adapted from Carvallo Ortiz (1988)

The early states

Among the pre-Columbian social groups, particular groups emerged as predominant, invading neighbouring spaces and imposing different socio-economic activities and cultural systems upon subordinated groups. The most significant societies in this respect within the continent prior to Iberian conquest were the Aztecs, the Incas and the Chibchas; the Mayas, Olmecs and Toltecs were also influential but had lost their authority by the fifteenth century (see Table 2.2, Figure 2.1).

Each of these societies extended its influence over a wide geographical area, establishing extensive administrative, economic and social systems for the organisation and control of these spaces (West, 1993). In the case of the Andean and Mesoamerican (Mexico and northern Central America) regions, these territorial empires followed one from another. In Mesoamerica, the Teotihuacán empire replaced the Olmec empire which was itself replaced by the Aztec empire.

Table 2.2 The emergence of pre-Columbian societies, 7000 BC–AD 1500

Archaic	7000–2500 BC	semi-nomadic tribes, early architecture and pottery, slash-and-burn agriculture
Pre-classic	2500 BC–AD 1	sedentary agrarian societies and economies, latterly small tribal states, development of pottery and weaving: Chavín (1000–500 BC)[a], Central Peru Olmec (1000–400 BC), Gulf of Mexico
Classic	AD 1–1000	stronger states: monarchic and theocratic, increased levels of social organisation, architectural development and city construction: Moche/Mochica, Nazca (200 BC–AD 600), Peru Maya (AD 300–900), SE Mexico, N Guatemala Tiahuanaco, Huari (AD 600–1000), Andes Teotihuacán (100 BC–AD 750), Mexico Zapotec and Mixtec (AD 300–1500), Mexico
Post-classic	AD 1000–1500	advanced organisation of state and society, militarised imperial expansionism, an absolute divine ruler – an aristocracy of warriors, priests and administrators – a massive underclass: Toltec (AD 900–1200), Central America/Mexico Chimu (AD 1000–1500), Peru Chibcha (AD 1200–1500), Colombia Tarascan (AD 1400–1500), Mexico Aztec (AD 1400–1500), Mexico/Central America Inca (AD 1400–1500), Andes

[a]Periods of greatest influence.
Source: Adapted from Williamson (1992)

The Inca empire provides a good example of the more complex pre-Columbian societies. The Incas had developed, through conquest, a regional empire that was of recent origin relative to the 20,000 years of Andean social history prior to the fifteenth century (Reid, 1984). Inca society was based on a system of state theocracy of sun worship and the Quechua language. The empire was founded upon the conquest of neighbouring agricultural societies that were predominantly feudal in structure, taking advantage of existing agricultural structures which became the sources of tribute to maintain the food security of the Incan urban centres.

The basic communal and productive units of non-Incan societies within the Andes at this time, *ayllu*, were grouped together under a local chieftain, a *kuraka*. The system functioned with the exchange of labour on the *kuraka*'s land in return for a part of the agricultural production and military protection; this system was known as the *mita*. Living and working in feudal communities enabled larger scale infrastructural projects to be initiated, such as irrigation systems or hillside terracing. Many of these paternalistic feudal features were maintained by the Incas as administrative communities within their expanding empire, for example, the *mita* became important in the exchange of labour for the redistribution of production.

At its peak, the population of the Inca empire could be estimated at 10–30 million people, the large majority of whom were non-Incas producing for their Inca overlords.[1] However, even before the *conquistadores* had arrived in numbers the empire was in decline. During the 1520s, the empire was

Plate 2.1 The Inca city of Machu Picchu, built in the Peruvian highlands. The *conquistadores* were unable to locate this important Inca centre.

Plate 2.2 The Mayan city of Chichen Itzá in Mexico's Yucatan region. In the foreground is an observatory, revealing the highly sophisticated nature of Mayan society.

suffering from a plague and when the ruler Huayna Kapac died, a war of succession followed between his sons, Atahualpa and Huascar. Atahualpa was victorious but the empire had lost its cohesion in the intervening period and many *kurakas* assisted the Spanish *conquistador* Pizarro and his small armed entourage when he marched on the Inca empire in 1532. The following year, Atahualpa was executed and Pizarro entered the Inca capital, Cuzco (Peru) (Hemming, 1970; Williamson, 1992).

The Inca empire was typical of the rise and fall of other pre-Columbian imperial societies. Either as a result of imperial overambition, whereby resources for administrative and control functions became overextended (Kennedy, 1987), or by internal divisions and civil strife, the pre-Columbian period witnessed the passing of numerous 'civilisations' with complex systems of territorial-administrative functions founded upon the subordination and disintegration of other social groups within their extensive imperial territories.

Plate 2.3 Participants in the Inca Festival of the Sun, which still takes place each year in Cuzco (Peru). Their language and dress reflect the indigenous culture and tradition that was in place prior to the Iberian conquest.

Plate 2.4 'El Trauco' is one of numerous folkloric idols on the island of Chiloé, southern Chile. Non-Christian beliefs and traditions persist despite the expansion of Catholicism in the continent. Traditional beliefs and symbols often became linked with those of the Catholic church, creating a quite distinctive form of Latin American Catholicism.

IBERIAN COLONIALISM

Conquest

The Iberian conquest of the Americas during the sixteenth century and the subsequent genocide (by armed massacre, disease transmission and harsh work regimes) of pre-Columbian peoples provided the context of spatial organisation of *Latin* America, originally named the kingdom of *Tierra Firme* by the Spanish. While the urban-rural divisions of the continent

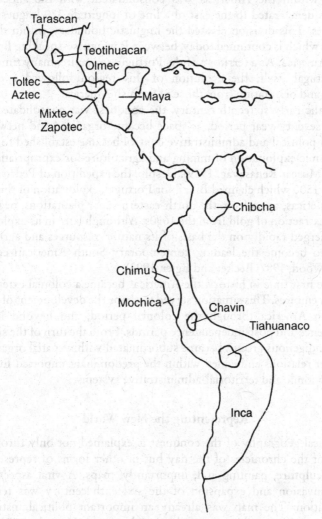

Figure 2.1 The pre-Columbian distribution of states
Source: Adapted from Glassner (1996)

became established by Iberian planning and production from the early sixteenth century, the broader divisions of the continent were also being drawn up.

The first and most significant act of Iberian political geography that shaped the evolution of the colonial Americas was the papal bull *Inter Caetera* of 1493 in which Alexander VI effectively partitioned the world into Old and New (Muscar Renasayag, 1989). The Treaty of Tordesillas between Spain and Portugal followed in 1494. By the treaty, Portuguese geopolitical presence within the Americas was consolidated, with the allocation of territories demarcated to the east of a line of longitude 370 leagues west of the Azores. This division created the linguistic boundary within the Latin Americas which is continued today, between Portuguese-speaking Brazil and Spanish America. As a territory of the Portuguese empire, many times larger than Portugal itself, the evolution of this region followed a different timetable and orientation from those of Spanish America.

From the early sixteenth century, the conquest was consolidated over a sixty- to seventy-year period, as space became organised and networks of centres of political and administrative control became established to oversee the economic exploitation of mining and agriculture for expropriation back to Spain (Muscar Renasayag, 1989). Despite the expedition of Pedro Alvarez Cabral in 1500 which claimed Brazil for Portugal, exploitation of this region of the Americas, apart from the north-eastern sugar plantations, began later with the extraction of gold from the 1690s. Although later in its exploitation, Brazil emerged rapidly on the basis of its natural resources and agricultural systems to become the leading contemporary South American economy (Russell-Wood, 1987; Becker and Egler, 1992).

For the first time in history, the Americas became a colonial extension of European empires. This situation served to shape the development of Central and South America through the colonial period, and beyond into the independence and post-independence periods. From the turn of the sixteenth century, indigenous groups became subordinated within spatial organisation and power relations embedded within the predominant imported European social, economic and territorial-administrative systems.

Representing the New World

The political geography of the conquest is explained not only through the medium of the chronicles of the day but in other forms of representation such as sculpture, painting and, importantly, maps. A vital aspect of the colonial invasion and expansion of the sixteenth century was territorial representation. The map was already an important political instrument, claiming to depict 'true' representations of the world (Mignolo, 1994). Maps from the period reveal key features of the geographical structuring of the 'New World' as constructed by the *conquistadores*, for example, the lack of

Plates 2.5 and 2.6 These monuments in the Parque Prado in Montevideo (Uruguay) commemorate the indigenous Indians of the region (Plate 2.5 above) and the European settlement of this 'frontier' (Plate 2.6 left). The tribes of this area were swiftly eradicated by the *conquistadores* and European expansion.

acknowledgement of the Amerindian presence within the continent. These maps reflect the ethnocentricity that served to legitimate the brutal acts carried out during the conquest and later colonial period. Eurocentrism and the ideologies of the Renaissance were at the heart of the philosophy behind 'discovery' and the 'civilising mission'.

In the late fifteenth century a Jesuit priest, Father Ricci, altered the European cartographic map of the world to place the Pacific at its centre. This resulted from a meeting with Chinese mandarins at his mission in Shaoking who were surprised at the position of China on the European-based map and the fact that the world was represented as a sphere (Mignolo, 1994). Ricci had recognised the ethnocentric conceptualisation of space inherent within the 'truthful' European representation of the world.

This ethnocentrism gives way to multiple ways of representing the world. Maps are full of meaning but they are representations of meaning, thus it is their interpretation that is significant rather than the image itself (Harley, 1992b). Interpretations of pre-Columbian and European maps reveal the cultural exchange in geographical knowledge that occurred during the sixteenth century. Brian Harley (1992a, 532) remarks that 'Maps can be shown to operate not only within material and practical processes but also as reified symbols of power, part of the political unconsciousness of European society.' In the case of the Americas, the *conquistadores* placed themselves, in Europe, in the centre of their maps. This is the omphalos syndrome, where ethnocentricity leads to mapping outwards from the navel (omphalos) or centre of the world, regarded by the ethnic group as its own perceived centre, not a geometric or arithmetic one. Ironically, the Incas also adopted this representative perspective. Cuzco, meaning 'navel of the world', was the centre of the Inca world (Sullivan, 1985).

Since pre-Columbian societies continued to predominate in particular regions of the Americas, especially in the Southern Cone, it was not until the late sixteenth century that representations of the whole of Central and South America were established. López de Velazco, from 1571 the official geographer and cosmographer of the Council of the Indies, constructed three regions within the Spanish territories: the Northern Indies, the Southern Indies, and the Western Indies (the Caribbean). Information, including a drawing of the locality, was requested by López de Velazco from public officials in all parts of the new colonies (Mignolo, 1994). This was the first systematic geographical survey of Latin America and was significant in that it was devoid of representations of Amerindian territoriality or organisation. The reason for this was that the information accumulated was not social geography, but essentially physical topography reflecting the nature and structure of settlement.

Mignolo (1994) makes the interesting point that this geographical information was written by Amerindian scribes in some places, and therefore the maps reflect their Amerindian styles of graphic representation of features and annotation. These maps document an alternative, non-European representation of the Americas at this time and are therefore important historical artefacts, reflecting a different perception and representation of space within pre-Columbian cultures.[2] This is especially significant in view of the general lack of pre-Columbian evidence of how space was organised, especially in a represented, visual form.

The mapping of the 'New World' by the *conquistadores* had the effect of shifting the perceived centre of the Old World from the Mediterranean (*medi-terra* – middle earth) to the Atlantic in order to account for the new continent. However, a shift of Latin America to the centre of a world map reveals its relationships within the Asia-Pacific region. Within the geopolitical and geoeconomic relations of the end of the twentieth century, the European-Atlantic bias of representation does not offer a good image of the developing connections of the Latin American continent.

Octavio Paz (1985), the Mexican writer, in his collection of essays *One Earth, Four or Five Worlds*, notes that predominant civilisations within world history have moved in an east to west direction over time, from the Middle East to Greco-Roman empires to the European empires. While this is merely a passing observation, the development of the Asia-Pacific political economies during the late twentieth century, particularly the USA and Japan, and the North American Free Trade Association (NAFTA), the Association of South-east Asian Nations (ASEAN) and the Asia-Pacific Economic Co-operation (APEC) regional blocs, give weight to the basis of such a concept. Clearly representational centres of the world are continuing to change. Ford and Hudson (1992) select the term 'New Middle Kingdom' to identify the conversion from an 'Atlanticist' view of geostrategic relations to an American-based zone of cooperation and competition, moving beyond the Eurocentric geopolitical world view of the Americas.

In terms of how the continent's political geography developed during the colonial period, a focus on urban and commercial networks provides an insight. López de Velazco was mapping the European territorialisation of the continent from its earliest stages. From small settlements based on maritime commerce and interior production sites of agriculture and mining, the Spanish and Portuguese began to develop networks of urban centres throughout their 'New World' empires. These networks incorporated settlements established prior to the conquest and acted as instruments of conquest in that they provided the framework of the colonisation process. They also centralised power and influence within the colonial Americas, a feature of development that continues into the twentieth century (Schaedel, 1978; Mauro, 1978). By the time Buenos Aires was founded for a second time in 1580 (having been abandoned in 1541) there were already approximately 250 urban centres within the colonial Americas.

The spread of these cities was from north to south, from contemporary Central America, Venezuela and Colombia to Peru, Ecuador, Bolivia, and lastly to the River Plate region. Brazilian development worked south from Olinda in the north-east to São Vicente in the south-east but experienced a slower urbanisation rate due to labour resource problems (Muscar Renasayag, 1989). In the rural areas, agriculture was carried out on landed estates using indigenous labour under harsh work regimes, the *encomienda* system. Under this system, indigenously controlled lands were distributed to *conquistadores* and colonists and the Indians would work the land to pay tribute to the Spanish monarchy.

The Amerindians were expected to be 'civilised' through Christian conversion while enslaved within this system. Within mining areas, particularly silver mining in Potosí, the Spanish jewel in the Americas, Amerindians were exposed to poor working conditions and received few benefits.[3] Only in the more inaccessible interior of Brazil and in peripheral areas of Spanish America, such as in Patagonia, did Amerindian groups escape insertion into this system of organisation and production, although they continued to be vulnerable to colonial expansion (Hemming, 1978).

By the mid-eighteenth century, the continent had been organised along relatively sophisticated lines in terms of administration and production activities. Pedro Murillo Velarde's (1752) *Geographia de America* (volume nine of his historical geography of the world) reveals the extent of the colonial geography of the continent. Murillo Velarde's volume followed two important works of this period regarding the Americas: *El Teatro Americano. Descripción General de los Reyes y Provincias de la Nueva España y sus jurisdicciones* by the Mexican José Antonio de Villaseñor y Sánchez (in two volumes, 1746 and 1748), and *Relación Histórica del Viage a la America Meridional hecho por Orden de S. Mag.* by two Spanish mariners, Jorge Juan and Antonio de Ulloa (two volumes, both in 1748).

These volumes revealed the opening up of the New World for geographical investigation during the central years of the 'Century of Light', the eighteenth century. These geographies focused on description, or illustration, rather than explanation. Murillo, a Jesuit, contrasted three geographical elements, topography, hydrography, and the human element he described as *corografía*, which was the description of the kingdom, the province or the region; the contemporary definition of chorography is the systemic description of spatial areas. Most certainly this was the contemporary basis of a political geography whereby landscape was linked with human activity and forms of organisation and authority.

Murillo's volume concentrated on the chorography of the Mexican and Peruvian areas of the continent because of his own personal knowledge of these areas and the fact that these areas were perceived as more advanced within the Spanish divisions of the continent, with political systems, laws, culture, cities, governments and officials. Apart from these areas, Latin America was divided, in his view, between groups governed by *caciques*, and the rest, barbarians: gentiles, idolators, and some atheists, as he described them. Also interesting is Murillo's identification of Geographical History, the title of the ten-volume series (1752, xxviii). He linked the importance of history (the accessory) to geography (the principal) in an understanding of the terrestrial world and the passing of the centuries. This demonstrates the importance of geography as a tool for understanding and representation during the colonial period.

The spatial organisation of *Tierra Firme*

The spatial consolidation of the colonial Americas differed according to region. In the case of contemporary Colombia, Mendoza Morales (1988–89) identifies that the region passed through three steps of territorial organisation: conquest and colony; the republican transition; and the establishment of the republic of Colombia. The region represented by contemporary Colombia was 'discovered' by Alonso de Ojeda in 1499 and ten years later two areas of colonial governance had been established within the kingdom of *Tierra Firme*: Uraba (or Nueva Andalucía), and Veragua (afterwards named Castilla de Oro). These governments were both dependent upon the royal audience based in Santo Domingo, in La Española (Hispaniola, now the Dominican Republic) from where much of Spain's New World was centrally controlled.

Under the rule of the *conquistadores*, political control and governance had the essential aims of organising and authorising further expeditions, in creating *capitanías* (captaincies) and in establishing a system of tributes and taxes, allocating land, and founding new urban centres (Mendoza Morales, 1988–89). A system of *encomiendas* was extended into rural areas, under the pretext of evangelisation of indigenous groups while they worked the land of *conquistadores*; meanwhile urban areas were organised as municipalities within which legislative codes were established. This dual division of urban and rural spatial organisation was used widely, initially having been constructed in contemporary Venezuela during the first half of the sixteenth century.

Spain's monarch Charles II (1516–1556) created the principal divisions of the kingdom of *Tierra Firme*, the royal audiences (*audiencias*) (see Figure 2.2). This was a first step towards a decentralised territorial-administrative system of the New World, away from the centralised structure emanating from Santo Domingo. Each audience had four controlling members (*oidores*) who would judge conflicts relating to civil and ecclesiastical matters, also issues relating to indigenous groups. Appeals in the case of the latter could only be heard in Spain by the Council of the Indies, the overarching administrators of the New World. Later conflicts between *oidores* led to the need for presidents who controlled the civil and military functions of the audiences. Central and South American territorial space was organised within the royal audiences of Panama (1538), Lima (1544), Quito (1563) and New Granada (1550).[4] As an example, the New Granada Audience was itself constituted by four territorial and administrative sub-divisions: the central territory of the New Kingdom of Granada and three

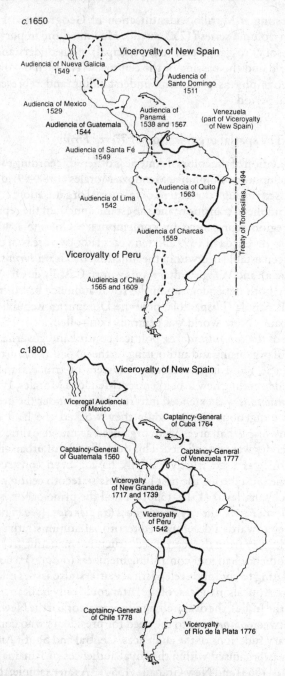

Figure 2.2 The development of colonial territorial-administrative structures
Source: Adapted from Williamson (1992)

peripheral areas governed from Santa Marta, Cartagena and Popayán (Mendoza Morales, 1988–89).

The shift from conquest to colonisation at the end of the sixteenth century led to different territorial organisation and controls from the early seventeenth century, directed by the new Bourbon monarchy in Spain. Under the Bourbons, the audience system of territorial administration became centralised under the vice-royalty system. In 1717, the Americas were divided into the vice-royalties of New Spain (Mexico and Central America), New Granada (northern Andes), New Castile (southern Andes), and the River Plate (established in 1776), with three further divisions: the captaincies of Cuba, Guatemala and Chile (Gwynne, 1994).[5]

In the Portuguese colony, the territorial division from the 1530s was also based on a captaincy system, established along horizontal divisions of the colonial territory. In terms of changes in authority, the *oidores* of the Spanish audiences were replaced by a provincial representative of the monarch, the Viceroy, who represented monarchical powers within the region. Internal divisions of the vice-royalties were organised into provinces, cantons and municipalities. The province and municipality system still predominates in late twentieth-century Latin America.

To use the New Granada Audience as an example of these changes, the audience became the Vice-Royalty of New Granada. The vice-royalty was comprised of fifteen provinces whose territorial unity the Viceroy was unable to enforce, in great part due to the geographical extent of the region, from the southern Amazon to the Caribbean sea to the north. The outcome was the establishment of numerous fragmented identities within the vice-royalty, such as pastoral, plains, island and coastal societies, and groups identifying more with particular urban centres rather than an all-encompassing governing entity (Mendoza Morales, 1988–89). This type of fragmentation sowed the seeds for the internal divisions of the vice-royalties more generally.

Events in Europe at the end of the eighteenth and in the early nineteenth century provided the opportunity for discontent in the organisation and control of the Americas to manifest itself. The independence struggles of the early nineteenth century led to the establishment of autonomous republics with their own territorial-administrative frameworks, replacing the larger geopolitical areas constructed by the Iberian colonial administrators. In the case of New Granada, a confederation of these new decentralised spaces, the Confederation of the United Provinces of New Granada, still maintained a connection with the colonial past.

During the colonisation period, particular regions became relatively more or less significant within Latin America. During the early period, the mines of the Andean region were most valuable to the Spanish crown. The administrative system utilised the traditional *mita* system to convert Amerindian peasants into miners. The Potosí silver mine became a symbol of the wealth generated within the Americas, and also of the brutal exploitation of expendable Amerindian labour. By 1600, Potosí had become the largest city in the Americas (Collier *et al.*, 1992). As an important port and nodal point of the most productive mining and textiles region in terms of the link with Spain, Lima remained the most significant administrative city within Spanish Latin America, reaching a population of 50,000 by the end of the eighteenth century (Reid, 1984). However, by this time a decline in mining output marked a shift in economic interest to other regions such as Venezuela and the River Plate area, for reasons of agricultural production.

Brazil suffered similar problems to Spanish America in establishing a coherent territorial-administrative structure. With coastal centres providing the main nodal points of imperial control over the interior, connecting the separate administrative *donatorios* (donatary-captaincies),[6] or colonial regions, a royal governor attempted to structure a more centralised territorial-administrative system than that existing in the Spanish Americas (Taylor Pittman, 1986) (see Figure 2.3). The movement of the Portuguese crown to Brazil during the Peninsular War (1808–14) against Napoleon was a further source of centralising influence within Portugal's most important colonial possession.[7] On King João VI's return to Portugal in 1820, Prince Pedro remained and secured a Luso-Brazilian independence from Portugal, declaring himself 'Constitutional Emperor and Perpetual Defender of Brazil' in 1822. While the Spanish American independence struggles terminated Spanish influence within Central and South America in the early nineteenth century, it was not until 1889 that the Portuguese monarchy submitted to pressure from landowners; Pedro II abdicated and the republic of Brazil was declared.

It was not only the organising influences of the imperial structures that were important in shaping the political geography of the Americas. In the south of the continent, the Jesuits were also influential in establishing evangelising missions to work with Amerindian groups. In the region of southern Brazil and Paraguay, Jesuit settlements shaped the organisation of Latin American space and the integration of Amerindians, principally the Guaraní, into European systems of belief, social organisation and production. By the early seventeenth century, the Jesuits had constructed thirty settlements totalling 200,000 inhabitants, but by the second half of the century they came into increasing conflict with the Spanish over the issue of Guaraní labour.

Following two Amerindian rebellions in opposition to Spanish interference in the region and the cessation of seven settlements to Portugal, the Jesuits were expelled from the regions and their settlements went into decline (Latin America Bureau, 1980). These areas came under the influence of the River Plate Vice-Royalty in the late eighteenth century but divisions in terms of predominant ethnic groups, and historical influences on the production system would emerge later in the post-independence struggles within the region. The opposition of Paraguayan peoples to Spanish interference from the River Plate region led to the proclamation of the Republic of Paraguay

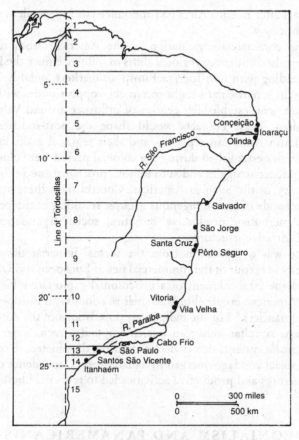

1 João de Barros e Aires da Cunha (Pará) 2nd part
2 Fernão Alvares de Andrade (Maranhão)
3 Antônio Cardoso de Barros (Piauí)
4 João de Barros e Aires da Cunha 1st part
5 Pero Lopes de Sousa (Itamaracá) 3rd part
6 Duarte Coelho (Pernambuco)
7 Francisco Pereira Coutinho (Bahia)
8 Jorge Figueiredo Correia (Ilhéus)
9 Pero do Campo Tourinho (Pôrto Seguro)
10 Vasco Fernandes Coutinho (Espírito Santo)
11 Pero de Goís (São Tomé)
12 Martim Afonso de Sousa (Rio de Janeiro) 2nd part
13 Pero Lopes de Sousa (Santo Amaro) 1st part
14 Martim Afonso de Sousa (São Vicente) 1st part
15 Pero Lopes de Sousa (Sant'Ana) 2nd part

Figure 2.3 Brazil's sixteenth-century donatary-captaincies
Source: Adapted from Johnson (1987)
Note: Grants were made to twelve captains with royal court connections; three captains received more than one grant. The grants were royal gifts or donations (hence donatary-captaincies).

47

in 1812, two years after Buenos Aires had announced its independence as the Argentine republic.

The changing commercial orientation of the Americas from the late sixteenth century also influenced regional shifts of influence since the Iberian colonies were trading with non-Iberian European markets, notably France and Britain. Britain in particular sought commercial opportunities within the south of the continent, establishing centres of influence around Valparaíso and Buenos Aires. These contacts would shape the post-independence orientation of Latin American republics and their regional make-up. The commercial activities established during the colonial period were continued after the independence struggles and autonomous proclamations in the early nineteenth century in the Spanish Americas. Control over these activities remained in the hands of non-indigenous groups, resulting in the perpetuation of non-Amerindian productive activities, social organisation and territorial-administrative structures.

Independence was a separation from the formal imperial ties of the Spanish monarchy in favour of the commercial ties of European and US trade and finance and the establishment of a *neo*colonial exploitative hierarchy based on Latin American creole élites and their relations with European and US traders and financiers.[8] For the indigenous populations of the Americas, independence was simultaneously an unchaining and a rechaining. While *mita* and *encomienda* systems slowly declined during the nineteenth century, new systems of social bondage emerged to replace them. Dependence on new land tenure structures and productive activities led to few civil liberties and opportunities.

NEOCOLONIALISM AND PANAMERICANISM

The struggles for independence

The anti-colonial independence struggles of the early nineteenth century were successful for three reasons: the Spanish were preoccupied in Europe with the campaigns against Napoleon; the territorial-administrative organisation of Spanish Latin America was fracturing along regional and local lines; and the leaders of the independence movement and their followers, linked with non-Iberian European sources of support, were more tenacious than the Spanish military.

With the forces of General Simón Bolívar, known as 'The Liberator', driving down from the north, and those of General José de San Martín working up from the River Plate region, the majority of Latin America had been liberated and had proclaimed independence by 1830 (see Lynch, 1973). With independence followed a new set of territorial-administrative structures of post-colonial spatial organisation which Luis Vitale (1992) describes as the 'balkanisation' of the Spanish Americas. While the new republics followed, to a great extent, the vice-royalty divisions of the colonial period, they sought alternative structures from which to develop liberal democratic systems of sovereign territorial states.[9] Within the constitutions of the republics, these alternative structures were assembled and legitimated.

Established by Bolívar, the largest republic was that of Gran Colombia founded in 1819. Gran Colombia was constructed by linking the Vice-Royalty of New Granada and the Captaincy of Venezuela, which was itself divided into the three departments of Venezuela, Cundinamarca and Quito. This new structure proved to be an unstable entity due to the diversity of regional influences and local power bases. In 1824, almost as a recognition of the Spanish errors of over-large territorial-administrative units, the republic became comprised of twelve departments with a total of thirty-eight provinces. Each department was controlled by a prefect and each province by a governor.

The framework of Gran Colombia linked the departments until Bolívar's death in 1830 when three nations emerged from the institutional breakdown of the state: New Granada (present-day Colombia), Venezuela, and Ecuador. In 1858, these three nations became confederated within the Republic of Colombia, later renamed the United States of Colombia. Although united, each state still had its own president, constitution, army and congress and could declare war on other states. In 1886 the republic returned to the structure of the 1819 departments with a constitution of centralism rather than federalism.

The Colombian case is typical of the organisation and reorganisation of the Americas following the declaration of independence. The Central American republics provide a similar example. The republics merged into the United Provinces of Central America in 1823, before fragmenting into their smaller territorial units, the present-day Central American republics, in 1838. Provinces, states, federations and confederations were the building blocks of the post-independence period, but there were considerable difficulties in establishing effective territorial-administrative units that were acceptable to regional interest groups and élite interest groups.

The most demanding interest groups were those of local political leaders (*caudillos*), often backed with a local armed militia, who were aligned with local landed élites and merchants to establish strong regional political and economic fiefdoms with large support bases.[10] Attempts to establish effective centralised systems with national-based identities consistently failed within this regionalisation of Latin American territorial space. Regionalisation under *caudillismo* created numerous obstacles to republican political author-ities which had to arbitrate and negotiate acceptable changes across sectoral divides.

Within the disarray of *caudillismo* and construction, deconstruction and reconstruction of spatial units and political-administrative mechanisms of control during the post-independence period, non-Iberian European and US

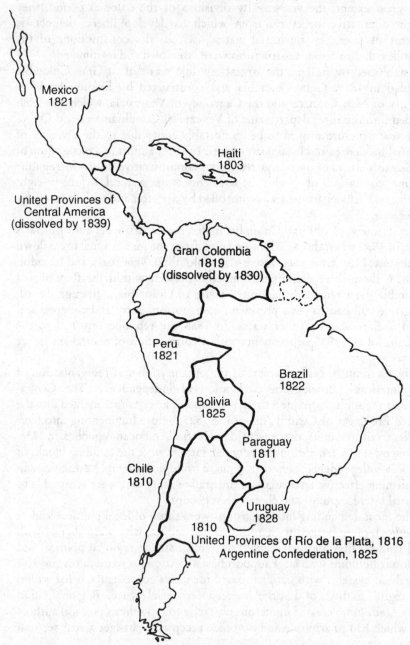

Figure 2.4 Latin America in the early nineteenth century (with dates of independence)
Source: Adapted from Williamson (1992)

commercial interests became critically important. Their importance within the development of the fledgling Latin American economies reached the point whereby, for example, they were involved in the formulation of early legislation relating to trade. It was the close integration of domestic élites with these foreign commercial élites that led to their strong influence in political and economic development strategies.

This commercial influence within nineteenth-century republican development meant that other geostrategic aims, relating to potential geographical advantages, could also be realised by merchants and financiers on behalf of foreign powers. Over one hundred years later, many of the geographical advantages that were secured are still contentious. The Falklands/Malvinas conflict is one such, dating back to the early post-independence period when British maritime trade was influential in the economic development of the River Plate region.

Box 2.1 The Falklands/Malvinas: the geopolitics of barren islands

In 1982, the war in the South Atlantic between Britain and Argentina characterised the difficulties associated with historically set, disputed territorial divisions within North–South geopolitics. It also revealed the complex relationship between militarism and domestic politics. The Falkland Islands/Malvinas are not fertile, agriculturally productive islands but are suitable only for sheep ranching, fishing and potential energy resources. Correspondingly they are not densely populated, with only 1,800 inhabitants in 1980. The reasons for the ongoing conflict between Argentina and Britain since the early nineteenth century lie in the islands' geostrategic location with respect to the South Atlantic, Antarctica and the Cape Horn link between the Atlantic and Pacific oceans. The March–July 1982 conflict over control of the islands, during which 1,000 lives were lost, was a late twentieth-century re-emergence of a historical dispute. The domestic politics of the two belligerent powers helps to explain the timing of the conflict and its rapid conclusion.

The history of the dispute dates back to Argentina's independence when Britain was at the zenith of its maritime supremacy and had already exerted its influence in the River Plate region by assisting Uruguay in its separation from Argentina. Only a brief conflict in 1806–7 prevented Britain from controlling Buenos Aires as well as Montevideo since the River Plate was an important region for Britain's geostrategic influence in the South Atlantic. In 1833, as part of its expansion in the region, the Falklands/Malvinas archipelago was seized from Argentina due to the location of the islands relative to the navigation routes of the Magellan Strait and Cape Horn. This 'Gibraltar in the South Atlantic' was defined by the British Lord of the Admiralty at the time as the key to the Pacific.

In 1843, when commercial whaling began in the South Atlantic, Britain's previous lack of interest beyond geostrategic possession altered and the islands were given political status as the Dependencies of the Falklands. They retained this title until 1962 when they were integrated into the British Antarctic Territory. Throughout this time, Argentina disputed British sovereignty over the islands. Argentinian claims are based on Spanish sovereignty over the South Atlantic, declared within the 1493 Papal Bull. Under the 1790 San Lorenzo Treaty between Spain and Britain, the Spanish possessions in the South Atlantic were transferred to Argentina. British claims date back to the 'discovery' of the islands in 1592 and 1594 by two British mariners, John Davis and Richard Hawkins. The events of the early 1980s marked the culmination of these counter-claims.

The conflict arose as a result of several factors. Most important was the ousting of General Viola under the pretext of illness in 1981, which brought General Galtieri to power. The armed forces were split by internal disagreements, having been in power since 1976, and the military's lack of political and economic direction brought criticism from both the Peronist and Radical parties. A further important factor was the human rights opposition to the military, seeking explanations for the thousands of *desaparecidos* missing since 1976. Galtieri attempted to counter opposition to the military by promoting a brand of populism within Argentina, targeting contentious foreign policy issues. One of these was the sovereignty of the three islands in the Beagle Channel over which Chile and Argentina had nearly gone to war in 1979. As in the Chilean case, the occupation of the Falklands/Malvinas in April 1982 was a distraction from Argentine political and economic difficulties and was an action guaranteed to engender populist support. Economic hardship and human rights abuses became obscured by nationalist fervour.

Britain's justification of its response to the Argentinian invasion of the islands on 2 April 1982 falls into the realms of the legal defence of British sovereign territory, a response to international aggression that threatened other dependent territories thus setting a precedent, and also the right of self-determination for Falkland Islanders to remain British. The outcome was a self-proclaimed victory for Margaret Thatcher, but it revealed the problematic nature of Anglo-US relations with regard to Latin America and also damaged British relations in the continent. For many countries, Brazil for instance, the British action was seen as an imperial gesture and Argentina gained support within the United Nations on the basis of the global North–South divide and the imperial legacy. The USA saw fit to hedge the Atlantic tie in favour of Panamericanism and the essence of the Monroe Doctrine, maintaining its commercial influence at all costs.

What the future may hold for the islands is difficult to predict, but the options may be independence with legislative and political autonomy for the islanders guaranteed by Britain (as in Cyprus), associated statehood with a monarch as figurehead but island affairs decided internally (bar foreign policy and defence), or self-government under some guise. Whatever direction is taken, Argentina's position is clear. President Carlos Menem declares openly that the Falklands/Malvinas will be under Argentine sovereignty once again by the year 2000.

Sources: Milia (1982), Honeywell and Pearce (1982), Agudelo Gómez (1988–89)

The role of European and US commercial and foreign policy within Latin America in the immediate post-independence period suggests that the new republics had not moved from a colonial to an autonomous political condition, rather they had discovered a middle ground. The republics, with their

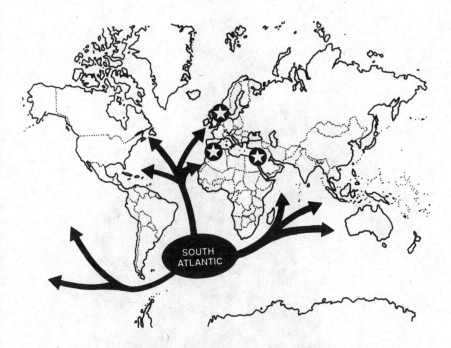

Figure 2.5 The geostrategic significance of the Falklands/Malvinas
Source: Horowitz and Sklar (1982)
Note: Stars mark the maritime gateways into the European 'heartland'.

reliance on foreign investment, loans and trade with Europe and the USA had become dependent along different lines from the formal empire linkages of the Iberian colonial period. No sooner had Latin America's new republics constituted themselves than they effectively began to replace their Iberian imperial ties with neocolonial ones. Rather than a new autonomous power geometry, the geometric relationship shifted across the Atlantic to North America for the northern Latin American republics, particularly Mexico, Venezuela and the Central American republics, while the Southern Cone became locked into northern, in preference to southern, European relations. These new ties were established via financial dependence and debt repayment obligations, and also within trading relations dominated by merchant houses (Greenhill, 1977). The ties established with the USA were along similar lines, although the US neocolonial tie had deeper political significance.

The USA and the Americas

In his 1823 State of the Union speech, the US President James Monroe established the future of the role of North America in the affairs of Central and South America. His speech, reacting to European overtones of

Plate 2.7 This monument of Columbus in Buenos Aires represents the continuing positive image of 'discovery' held by the more European sectors of Latin American societies. The Iberian conquest also marked the beginnings of intercontinental military geopolitics in Latin America.

Plate 2.8 The continuation of intercontinental military geopolitics 500 years later is symbolised by this statue commemorating the Argentinians who died in the Falklands/Malvinas War. Buenos Aires.

intervention in the independence struggles within Spain's empire, declared that the Americas should not be considered within the colonial ambitions of the European states and that, in return, US foreign policy would respect European colonial possessions elsewhere and not intervene. The speech has had far-reaching implications for relations within the Americas ever since, as it established the starting point for new declarations of US 'rights' to the Americas which have been central to the US concept of its 'manifest destiny' to control the Americas and profit from these southern frontier lands.

The most significant additions to the basis of the doctrine were advanced by Secretary of State Richard Olney in 1895 and President Theodore Roosevelt in 1904. The Olney Corollary of 1895 was added during the border dispute between Venezuela and British Guiana when Olney declared that the USA was practically sovereign in the continent. Based on its 'manifest destiny' within the region, US policy was justified in its extended control over the 'Other' Americas. Issues of sovereignty and territoriality were subsumed within this.

The Roosevelt Corollary of 1904 extended the doctrine to a declaration of US intervention within Central and South America in cases of weakened political systems leading to civil unrest, and also cases of 'chronic wrongdoing'. Intervention in Panama in order to maintain commercial links between Atlantic and Pacific coasts, via control of the Panama Canal and its parallel territorial zone, provides the most significant example of US influence during the late nineteenth century when the Central American and Caribbean regions, the US 'backyard', provided the focus for interventionism.

While these two corollaries to the original doctrine consolidated the foundations of US policy within the Americas, the essential elements of US 'manifest destiny' within the Americas, such as its right to oversee and intervene in Latin American republics' affairs, became the justification for its political and economic activities. This continues to the present day and has been invoked on repeated occasions within the post-Iberian, neocolonial period. Without doubt, a new power relationship was established from 1823 within the Americas. This new relationship was, and is, not one of direct political control, governance and systematic exploitation, rather a more subtle campaign of indirect influence and interventionism that may be defined as neocolonialism since there are important unresolved issues of political autonomy and sovereign status. From the US role in the independence process of the early nineteenth century to the Panama invasion of 1989, the exercise of these inter-Americas (North–Central, North–South) neocolonial ties has been unremitting.

During the twentieth century, US policy has emerged in three distinct phases: the 'good neighbor' policy of the 1930s, the Alliance for Progress of the 1960s, and the low intensity conflict of the 1980s. These phases have provided the strategies behind US neocolonial relationships. The period from

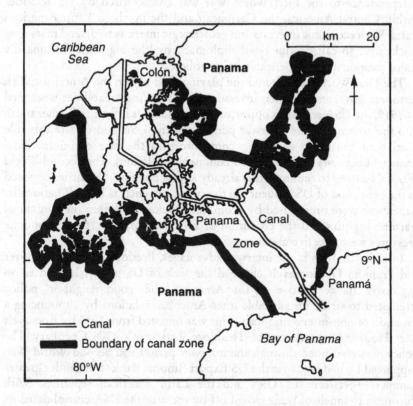

1846 USA gains right to free transit across Panama under the Mallarino–Bidlack Treaty between the USA and New Granada
1850 US–British Clayton–Bulwer Treaty whereby each state agrees not to gain exclusive control over any Central American canal
1880 Failure of sea-level ship canal engineered by the Frenchman De Lesseps
1901 Hay–Pauncefote Treaty replaces Clayton–Bulwer Treaty and gives exclusive US political control over an interoceanic canal built and defended by the USA
1902 The Spooner Law gives US President Theodore Roosevelt support for a US-built, US-defended canal
1903 The state of Panama created and the Hay–Bunau Treaty signed with the USA. A US concession over the canal route granted in perpetuity. The Canal Zone is established, 6–10 miles either side of the route
1977 Carter–Torrijos Treaty restores the Canal Zone to Panamanian sovereignty from 1999

Figure 2.6 The Canal and US–Panamanian relations
Source: Adapted from Girot (1994)

independence to the First World War was characterised by US activities within Central America, the Caribbean and the northern Latin American states. With economic interests and geostrategic interests to defend in its own 'backyard', so-called 'gun boat' diplomacy or 'the big stick' became the major weapon in US American foreign policy.

The First World War altered the playing field within the Americas. With European governments, their investments and trading capabilities weakened by 1919, the USA seized its opportunity to extend its influence further south into the continent. The interwar period saw the USA consolidate this role relative to European economic competitors in the face of international financial insecurity and depressed commercial relations. The Second World War only served to reinforce this already strong position. Countries opposed to the expansion of US influence in the continent, such as Perón's nationalist Argentina, were pressurised by economic blockades. Within the context of wartime logistics and the condition of European markets, these economic pressures were effective measures.

In order to press home interwar advantages, Presidents Herbert Hoover and Franklin D. Roosevelt changed the tack of US policy. Instead of an aggressive stance relative to the Americas, the 'good neighbor' policy attempted to strike up amicable inter-American relations by announcing a situation of non-interventionism. This was initiated from 1933 by Roosevelt after Hoover's amendment of Theodore Roosevelt's 1904 Corollary. The policy was continued through the interwar period and Second World War, supported by tied loans via the US Export–Import Bank to engender greater commerce between the USA and the Latin American republics. With European channels of trade closed off by the war, the USA consolidated its trading position within the Americas.

The Organisation of American States (OAS) was also an important mechanism for creating more pro-USA interdependency within the Americas. The OAS came into being in 1948 at a meeting in Bogotá, redefining the ineffectual International Union of American Republics established in 1890 (Collier *et al.*, 1992). The timing of setting up the OAS was important since it followed the 1947 Rio de Janeiro Panamerican defence agreement, the Rio Treaty, which established the collective security of the Americas through joint action against aggressive states within and beyond the Americas. The Charter of the OAS (1948) and the American Treaty on Pacific Settlement (1948) followed swiftly (Glassner, 1996).

The role of the USA in the Organisation and as the key player in any Americas-wide defensive action ensured the maintenance of its influence. Countries seeking aid and economic advantages have sided with the USA within the OAS to ensure good favour. For example, during the 1980s, Ecuador had the reputation of being Washington's most reliable ally under President Febres Cordero, combating the debt crisis with harsh austerity as advised, and blocking the OAS resolution to condemn the Pinochet dictatorship. Ecuador, by following the US Panamerican line and neoliberal model, earned President Febres the acclaim of having instituted a form of 'Andean Thatcherism' (Corkill and Cubitt, 1988).

The Cuban revolution that culminated in a communist victory in 1959 served to shake the Panamerican commercial and defensive union within which US influence predominated. The US response was swift, acting through the OAS to impose an inter-American trade embargo on Cuba, and initiating the unsuccessful Bay of Pigs invasion in an attempt to restore pro-US authority over the island. To confront the Cuban revolutionary rhetoric of continental revolution in the Americas, at a time when US military attention was directed towards Vietnam, President Kennedy initiated the Alliance for Progress in 1961. US economic aid programmes of the Alliance for Progress, totalling $20 billion over ten years, were put in place throughout Central and South America with the aim of countering the activities of communist movements (Skidmore and Smith, 1989); representatives of Latin American administrations signed up to the programmes at an OAS meeting in August 1961. The failure of Che Guevara's Cuban guerrillas to initiate revolutionary activities on the Latin American mainland, and the betrayal by communists in Bolivia that led to his death, suggests that the Alliance for Progress succeeded in its ideological aims.

The social and economic development associated with the Alliance programmes were supplementary to the ideological foundation of aid-led intervention since the policies of counter-insurgency were foremost within the minds of US state officials. The Alliance for Progress was essentially an anti-communist programme. The military training provision and arms equipment supplies to Latin America in the post-1960 period bear witness to this. These military programmes were integral to the equipping and training of the military forces that assumed power over the next ten years: in Brazil in 1964, Uruguay and Chile in 1973, Argentina in 1976. There is little doubt that such forms of authoritarianism provided the stable regime types desirable for US commercial activities and interests within the Americas.

These forms of intervention of the 1960s were markedly different from those utilised during the 1980s in order to counter the threat of the Nicaraguan revolution (1979). As with Cuba, Nicaragua was isolated from its trade partners by an embargo while opposition to the revolution was financed and trained. The mechanisms by which the USA adopted indirect opposition to movements or governments opposed to 'Yankee imperialism' within Central America was the policy of low intensity conflict (Holden, 1993). By financing, training and arming opposition to destabilising forces, the US government maintained wars of attrition within its backyard without the direct involvement of US armed forces. This policy served to maintain the instability of the Central American isthmus throughout the 1980s, leading to the period being referred to as the 'lost decade'. Only towards the end of the 1980s did negotiations begin between the belligerent forces within the

Table 2.3 US neocolonialism in Central and South America, 1823–1994

1823	Monroe Doctrine announced
1848	Texas ceded by Mexico to USA
1895	Olney Corollary to Monroe Doctrine, recognising practical sovereignty of USA over the Americas
1898–1902	USA occupies Cuba after independence from Spain. The Platt Amendment (1901) gives USA right of intervention in Cuba
1903	Panama independence assisted. Canal construction begins in 1904
1904	Roosevelt Corollary to Monroe Doctrine, promoting intervention
1906	US governor replaces Cuban president
1912–16	US armed forces occupy Nicaragua (brief absence, 1925). US establishes rights to build canal through Nicaragua (1916)
1914–18	First World War: increased US trade, investment and influence within the Americas
1915–34	US invasion and occupation of Haiti
1928	Clark Memorandum on the Monroe Doctrine promoting non-intervention
1939–45	Second World War: further consolidation of US trade, investment and influence within the Americas
1947	The Truman Doctrine to defend peoples everywhere against communism. A clear message to Latin America, justifying US intervention
	The Rio Treaty (Inter-American Treaty on Reciprocal Assistance), a regional security pact of the Americas signed. USA plays important role
1954	CIA-backed coup in Guatemala
1961	US trade embargo on Cuba following the revolution (1959). A US-backed counter-revolution ends in the Bay of Pigs fiasco
1962	The Cuban 'missile crisis' ends with USSR withdrawing weapons in exchange for assurance of US promise not to invade
1965	US troops enter Dominican Republic to suppress democratic protest movement
	Pro-US president elected the following year (previous US involvement in the Dominican Republic: occupation 1903–04; invasion 1905; occupation 1916–24)
1970–73	Undermining of the Allende government in Chile. Pinochet dictatorship recognised immediately after coup in 1973
1980s	Low-intensity conflict supported by USA in Nicaragua, El Salvador, Guatemala
1981	USA ends aid to Nicaragua following revolution (1979). Contra war begins
1989	Invasion of Panama to capture President Manuel Noriega for drugs activities and non-cooperation with US policies
1990	Bush's 'Initiative for the Americas' for closer integration, intensification of the 'drugs war', and the proclaimed defence of democracy and human rights
1994	USA asserts role in the Caribbean basin via military intervention in Haiti

Sources: Gunson *et al.* (1989), Wynia (1990)

Table 2.4 The Declaration to the Peoples of America (the objectives of the Alliance for Progress), Punta del Este (Uruguay), August 1961

1	To improve and strengthen democratic institutions
2	To accelerate economic and social development
3	To carry out urban and rural housing programmes
4	To encourage, in accordance with the characteristics of each country, programmes of agrarian reform
5	To assure fair wages and satisfactory working conditions for all workers
6	To wipe out illiteracy and increase facilities for secondary and higher education
7	To press forward with programmes of health and sanitation
8	To reform tax laws, demanding more from those who have most
9	To maintain monetary and fiscal policies which will protect purchasing power
10	To stimulate private enterprise in order to encourage economic development

Source: James (1969)

Central American conflicts. The Esquipulas Summits from 1987 led to increased stability and have secured more peaceful paths of development during the 1990s (Dunkerley, 1994).

For the 1990s, following OAS opposition to the 1989 US invasion of Panama to remove General Noriega, the US strategy has once again been one of apparent good neighbourliness. President Bush's declaration of a new orientation towards the isthmus came about with the 'Initiative for the Americas' of November 1990. The initiative is a clear indication of US aspirations within the continent, as police force and social worker (Carpenter, 1991; O Tuathail, 1992). The USA pledged $300 million to be administered by the Bank of International Development in this initiative, which centred on commerce, investment and debt.

The effect of the finance has been increased US involvement in the region via its commitment to drug traffic control and in its proclaimed defence of democracy and human rights as determined by the US State Department.[11] In terms of US assistance via normal financing channels, US state support for Central America has diminished since 1989. However, private economic interests still ensure the US presence.[12] Horacio Godoy (1993, 65) raises the astute point that this was clearly an Initiative *for* the Americas rather than *of* the Americas. Despite such scepticism, the 1990s have marked an end of aggressive, armed intervention, direct or indirect, within the continent, as more attention has been focused on economic initiatives within regional blocs, such as NAFTA, and also in the maintenance and deepening of the neoliberal model within the Americas.[13]

Post-Cold War US–Latin American relations

The end of the Cold War, which provided the context for US activities within the continent during the 1947–90 period, provides alternative directions for

US intervention and the exercise of power over the Americas. The Panama invasion, typical of US 'big stick' actions in Central America dating back to its nineteenth-century 'gunboat diplomacy', met with vociferous opposition from the other members of the OAS. Clearly the post-Cold War world offers the opportunity for the emergence of different power geometries within the geopolitical framework of the Americas. Richard Ullman (1994, 13) describes the Panamericanism of the Cold War as 'both a glue and a caustic. It served to bond the northern and southern parts of the hemisphere together, but it did so in a way that burned.' The question is to what extent this type of criticism will continue to hold fast during the post-Cold War period.

There are opportunities for interconnectedness without the degree of manipulation and brutal antagonism that characterised Cold War US policy in the Americas (Holden, 1993). These opportunities appear to be increasingly within the realms of the geoeconomic, although the experience of debt management within the framework of neoliberalism since the early 1980s has not been an example of the realisation of such opportunities. Power has made a post-Cold War shift from the 'big stick' to the open market. It is within continental neoliberalism that the USA seeks the opportunity to exert its influence throughout the Americas, via investment and commerce. With less direct influence over Latin American political systems due to the almost omnipresent neoliberal model, there is also the opportunity that neoliberal relations between the USA and Latin America will replace the neocolonial ones that have characterised the 150 post-independence years since the Monroe Doctrine.

CONTEMPORARY CONTINENTAL GEOPOLITICS

Military geopolitics

While US-dominated Panamericanism has been most influential in the post-independence development of Latin American political economies, the continent has also been wracked by internal divisions and disputes which have also had destabilising influences on the Latin American inter-state system, the construction of unifying frameworks for economic stability and the promotion of intra-continental commerce and investment. Many of the disputes are based on boundary delineations and the consolidation of frontier regions which have low population densities and are contested. While relatively few of the disputes have led to armed conflict, there are so many that inevitably there have been military engagements.

It is true to say that military authoritarian regimes have raised these boundary disputes more than democratic administrations due to the Latin American tradition of geopolitics in military academies. Escolar et al. (1993) argue that this situation arises as a result of the stressing of the link between nationalism and territory, the 'patriotic territory', within public education and culture. This notion is central to military geopolitical thinking and is perpetuated within military institutions such as geopolitical institutes and training colleges (Escolar *et al.*, 1993, 362):

> The territorial mythology has fostered pseudo-scientific arguments such as those of the 'geopolitical experts', for whom the territorial myth is the only foundation of discourse. The discourse has remained undisputed in spite of its evident irrationality.

The Latin American geographical tradition is closely aligned with military geostrategy (Child, 1985; Morris, 1996). The best example in support of this statement is the predominance throughout the continent of military geography institutes. These are often the centres of national cartography and geopolitical academic and military training. It is this close association of the military with geopolitics that leads to the constant struggle for territorial gain and geopolitical advantage. Militarism and cartography, principally used for political propaganda purposes, are the footings of the political geography foundations that exist (Dodds, 1993b). This traditional geopolitics is based on formulations originating from late nineteenth- and early twentieth-century thought which emanated from a transfer of traditional European geopolitical ideas.

Much of the explanation for the importance and sensitivity of boundary issues is this geopolitical tradition within Latin American military academies and the emphasis that the relatively youthful territorial units of the republics have placed on the preservation of territorial integrity and the development of claims. These claims are based on historical precedent at various junctures of the first half of the nineteenth century when the republics were attempting to establish recognised state boundaries and the effective organisation of national spaces for administrative purposes. For example, the loss of Panama from the Colombian republic in 1903 was merely a further blow to the territorial integrity of the state since Colombia has lost half of its 1810 *uti possidetis juris* territory via international negotiation since independence (Mendoza Morales, 1988–89). The *uti possidetis juris* is the most important basis for Latin American territorial disputes. It is based on the 1819 Congress of Angostura declaration, when Bolívar, the president of the congress, announced that the new states should maintain the same boundaries as the colonial political-administrative territorial units of 1810 (Astudillo Espinosa, 1993).

Contemporary disputes

Apart from the Anglo-European relativism of Derwent Whittlesey's comment that Latin America is geopolitically immature, it is true to say that Latin America, as a continent of relatively new territorial and sovereign nation-states, has a large number of ongoing territorial disputes. The continent is

Table 2.5 Latin American conflicts by salience and persistence[a]

Salient (high levels of tension and actual or potential hostilities)		
Argentina–UK	territorial	persistent
Cuba–USA	ideological	recent
Nicaragua–USA	ideological	recent
Nicaragua–Honduras	ideological	recent
Nicaragua–El Salvador	border, ideological	recent
Nicaragua–Costa Rica	border, ideological	recent
Potentially salient (grievance high, little immediate prospects of an outbreak of hostilities)		
Ecuador–Peru	border	persistent
El Salvador–Honduras	border, migration	recent
Guatemala–Belize	territorial	persistent
Controlled (likelihood of conflict low, tensions ephemeral due to power disparities of states)		
Colombia–Nicaragua	territorial	persistent
Cuba–Jamaica	ideological	recent
Argentina–Brazil	hegemonic, resource	persistent
Bolivia–Chile	territorial	persistent
Peru–Chile	territorial	persistent
Haiti–Dominican Republic	territorial	persistent
Mexico–USA	migration, resource	persistent
Colombia–Venezuela	migration, resource	persistent
Argentina–Paraguay	migration, resource	persistent
Argentina–Bolivia	migration, resource	persistent
Bolivia–Brazil	migration, resource	persistent
Brazil–Guyana	migration, resource	persistent
Brazil–Paraguay	resource, migration	persistent
Chile–UK	Antarctic claim	persistent
Argentina–UK	Antarctic claim	persistent
Trinidad–Venezuela	territorial	persistent

[a]Obviously overlap occurs between these categories and salience will alter over time. Ideological tensions between distant states have been excluded.
Source: Little (1987)

still undergoing the evolutionary process of the establishment of territorial integrity, in the face of historical precedents of territorial occupation and control. Among the numerous recent geopolitical confrontations, the examples of the Chile–Argentina boundary dispute, the Ecuador–Peru territorial division, and the Bolivian access to the sea, establish an understanding of the complexity of geopolitical disputes and their resolution.

Plate 2.9 Unlike many of the disputed borders within Latin America, the Iguazú Falls which divide Argentina, Paraguay and Brazil provide a strong physical demarcation.

Box 2.2 Landlocked: the Bolivian case

The sea, as a resource and as the medium of maritime trade, is a critical component of national political and economic development. With two landlocked states, Latin America provides prominent examples of the significance of the sea for commercial purposes and as a source of economic activity. Paraguay and Bolivia, as low income and low social indicator countries, exhibit the difficulties associated with the condition of being landlocked. It would not be too bold a statement to link the impoverished nature of Paraguay and Bolivia's socio-economies to their landlocked status. While Switzerland provides the contrasting experience, its financial sector niche has ensured a different development path.

For Bolivia, the struggle to regain an access to the sea since it lost its land corridor to the Pacific at the end of the War of the Pacific (1879–82) has been its principal geopolitical objective. General García Meza, President of Bolivia in 1980, announced that 'We must enable ourselves to reconquer our sea with armed force.' Such declarations diverted contemporary Chilean preoccupations with Argentina.

Under the delayed 1904 Treaty of Peace, Friendship and Commerce ending the War of the Pacific, Bolivia ceded back to Chile any rightful ownership of sea access that had been established in the colonial period. In return, Bolivia received the right to utilise the ports of Arica and Antofagasta; these rights were extended in 1937 and 1955. Arica became a key port for three countries and was considered the maritime entry into the South American 'heartland', an important geostrategic location. Maritime trade was also made possible through the Peruvian ports of Ilo, Mollendo and Matarani. Despite these gestures, territorial rights to the sea have continued to preoccupy Bolivian administrations and geopoliticians.

From the setting up of UNCTAD's Committee on Preparation of a Draft Convention Relating to Transit Trade of Land-locked Countries in 1964, and the 1965 UN Conference on Transit Trade of Land-locked Countries, the situation of landlocked states has become increasingly recognised. A breakthrough came from the UN Convention of the Law of the Sea (UNCLOS III) in 1982 which witnessed the world's landlocked states acting as a bloc to establish access to the sea and free transit to and from the sea. While the Convention does not guarantee these, international acceptance of their claims has put increasing pressure on those states whose territories lie between them and the sea. The most negative outcome of the Convention for landlocked states was the 200-nautical mile exclusive economic zone (EEZ) established off nation-state coastlines. The resources within these EEZs have been largely closed off from landlocked state exploitation. Nevertheless, the outcome of the Convention and the drive towards sea access rights for Bolivia have been successful in the years following UNCLOS III.

In 1987, Chile, Peru and Bolivia re-established talks regarding Bolivian access that had failed in the mid-1950s and mid-1970s at a presidential level. In the mid-1970s talks, a corridor had been agreed by Generals Banza and Pinochet only to be turned down by Peru, which favoured the internationalisation of Arica; this led to belligerent Bolivian rhetoric. Recent developments have seen the cessation of a Free Zone to Bolivia in the port of Ilo in 1992, and an Economic Complementation Agreement between Chile and Bolivia signed in 1993. While Bolivia still does not have sovereign rights over a territorial access to the Pacific, the economics underlying the need for maritime access have, for the present, been alleviated.

Sources: Ghisolfo Araya (1987), Pinochet de la Barra (1987), Aravena Ricardi (1988), Glassner (1993), Gangas and Santis (1994)

As the leading powers within the Southern Cone, Chile and Argentina have maintained poor diplomatic relations since independence in an attempt to establish versions of a geo-history of sovereignty in the region (María Miro, 1985). Much of the animosity between the two nation-states has been derived from failures to reach agreement on the fixing of boundaries between the two national territories along the Andean frontier and within southern

Patagonia, particularly the navigable waterway known as the Beagle Channel (Mateo Martinic, 1981). In 1979, with both states controlled by military authoritarian regimes, the two sides almost went to war over the dispute. The mediation of the Pope later resolved the conflict through the declaration of a Treaty of Peace and Friendship.

In 1985, Argentina had returned to democracy but was still suffering from the outcome of the Falklands/Malvinas conflict and a high degree of military opposition from within civil society. The signing of a treaty with Chile accepting the conditions of a former treaty in 1977 was the result. The 1985 treaty agreement followed a long line of accords dating back to 1855 when both sides agreed to the *uti possidetis* of 1810 when independence was proclaimed by the United Provinces of the River Plate. A new boundary treaty was established in 1881 and Tierra del Fuego was divided in 1895. This treaty was superseded by a general arbitration treaty in 1902 before a further independent arbitration by a British committee in 1977 which led to the 1979 confrontation. This history reveals that the geopolitical issue of boundaries has persistently raised its head in post-independence history (*Revista Geográfica de Chile 'Terra Australis'*, 1984–85). The same is true throughout the continent.

Another common theme in Chilean and Argentinian geopolitics has been the significance of maritime geopolitics, or political oceanography as Riesco Jaramillo terms it.[14] While Argentina has centred its maritime geopolitics on control of the South Atlantic and access routes to the South Pacific, Chile has attempted to establish its control over the South-east Pacific and routes to the South Atlantic. In the Chilean case, due to restricted territorial space, its maritime heritage and an oceanic destiny have been stressed, based on the hispanic 'discovery' by sea (Arnello Romo, 1988a, 1988b; Martínez Baeza, 1988).[15]

Islands play an important part in this tradition of maritime geopolitics, most clearly in the Argentine–Malvinas case. Within the Ecuadorean maritime consciousness, the Galapagos Islands are considered geopolitically significant; Easter Island performs the same function in Chilean geopolitical consciousness. Both of these examples reflect the problematic balance between the islands' commercial value as tourist attractions and the difficulties in establishing such distant territorial units in the national geopolitical consciousness (Hidrovo Penaherrera, 1985; Ghisolfo Araya, 1985). It is with the utilisation of island possessions that both Chile and Argentina have established zones of geopolitical influence that are inclusive of terrestrial and maritime spaces (Morris, 1996). Riesco Jaramillo (1985, 26), a Chilean geopolitician, cloaks this strategy in the familiar terms of Halford Mackinder:

> Whosoever controls the waters of the South Pacific Ocean projects geopolitical control over the west coast of South America. Whosoever controls the South Pacific Ocean simultaneously exercises control over the Antarctic continent.

Within Chile, geopolitical consciousness is raised by extending the land-based Chile to the west and the south. As the Pacific increases in global geopolitical significance, through Asia Pacific Economic Co-operation (APEC) for example, Mario Arnello Romo (1988b) projects a tri-dimensional vision of Chile – South American, oceanic (encompassing Easter Island and Sala and Gómez islands) and the Antarctic. This geostrategic role accentuates Chile's *austral-magallánica* (southern-Magellan Straits) position, dominating access to the Atlantic and Antarctic. This triangular notion of Chilean national space, projecting west to Easter Island, north to Arica and south to Cape Horn and Antarctica, contrasts with the commonly conceived conception of Chilean territoriality of a narrow stretch of land, half desert, squeezed between the Andean mountain range (the *cordillera*) and the Pacific. A similar triangular concept of geopolitical space has been employed in Argentina to encompass the Falkland Islands/Malvinas (see Morris, 1996).

Argentina and Chile have sought to extend their geopolitical influence over large areas of maritime space, but Paraguay has been unable to. As a landlocked state, Paraguay has suffered from a succession of post-independence struggles that has left it weak and significantly diminished in terms of national territory. The evolution of Paraguayan territoriality since independence provides a strong example of the significance of terrestrial boundary contestation within the continent. From independence in 1812, much to the chagrin of the River Plate regional centre in Buenos Aires, Paraguay established itself as an agricultural producer, initially focusing on livestock and the growing of *yerba máte*, used widely as a hot infused drink.

While clashes with Portuguese armed forces on the northern border were commonplace and Brazil provided the major threat to Argentinian predominance in the region, the new republic of Argentina refused to recognise or trade with Paraguay because of its early independence breakaway (Latin America Bureau, 1980). With agricultural self-sufficiency attained by the mid-nineteenth century, Paraguayan development passed through its 'golden years' in 1842–62. Its development strategy contrasted with the foreign merchant involvement with the fledgling republics along the Atlantic and Pacific coasts. The state was interventionist and successful, controlling exports of *máte*, timber, livestock, cotton and tobacco, and developing infrastructural projects, such as rail connections and telegraph lines.

Paraguay's successes led to anxiety in Argentina and Brazil regarding their provinces flanking Paraguay, such as Brazil's south-western region of Matto Grosso. When Brazil intervened in Uruguay to overthrow the ruling Blanco party in 1864, Paraguay assisted Uruguay, since its neighbour provided an important distraction for competing Brazilian and Argentinian interests. Because Paraguayan troops were refused entry to Uruguay via Argentina's territory, Paraguay declared war on Argentina as it had done on Brazil. In 1865, Brazil supported a favourable Colorado party administration in

Uruguay and, with the support of Argentina, the three countries formed a Triple Alliance against Paraguay with the intention of annexing its territories. The outcome for Paraguay was devastating as the 1865 population dropped from 500,000 to 250,000, with only 28,000 male Paraguayans remaining after the war. A quarter of Paraguayan territory was annexed and divided between Argentina (60 per cent) and Brazil (40 per cent) (Latin America Bureau, 1980). Paraguay has yet to recover from the effects of the Triple Alliance War.

Immediately after the conflict, the Paraguayan state was forced to sell off much of the republic's land and other national assets. The Chaco War which broke out in July 1932 was a further blow to the Paraguayan state and its territorial integrity. In 1928, conflicts between Paraguayan and Bolivian forces broke out in the Chaco region of the country to the north as a result of competing claims to oil resources and an attempt to establish a Bolivian sea route to the Atlantic (for Standard Oil which monopolised the Bolivian industry), the Pacific route having been lost in the War of the Pacific (1879–82). The war that ensued was finally won by Paraguay in 1935 after further devastating losses in terms of life, territory and damage to the Paraguayan economy.

Despite victory, there were to be further complications for Paraguay. When the Paraguayan government attempted to gain control over the Chaco for its land reform policy, Standard Oil appealed to Argentina for support. Spruille Braden, the US Ambassador to the Chaco Peace Conference, was a key figure in the conference which decided the cessation of an area of the Chaco (east of Villazón). A treaty was signed in 1938 and the seceded area became the zone of the Bolivian petroleum boom of later years. The Paraguayan victory had been a Pyrrhic one. The political turmoil that followed led to a civil war in 1947 from which President Stroessner emerged in 1954 to dictate Paraguayan development for the following thirty-five years.

The most contemporary ongoing dispute in Latin America that has given rise to armed conflict is that between Peru and Ecuador, once again a conflict over the early nineteenth-century post-independence delineation of territories. Ecuador was absorbed into Bolívar's Gran Colombia in 1822 but this state of affairs conflicted with Peruvian claims to southern Ecuador. The claim led to Peru and Colombia going to war in 1828–29, and to Peru's defeat.

In 1842, Ecuador, by then a republican state, claimed the Peruvian territories of Jaén and Maynas. War was only averted because of a government collapse in Ecuador. The dispute has since simmered among claims and counter-claims. Other boundaries that Ecuador has contested are those with Colombia based on the colonial period delineations. Ecuador only relinquished these claims after conflicts in 1830–34 and 1841–42. The 1916 Muñoz Vernaza-Suárez Treaty settled the issue, but Ecuador's northern border with Colombia remains undeveloped and a drug trafficking zone (Corkill and Cubitt, 1988; García Negrete, 1993). On the maritime front,

69

Ecuadorean control over the Galapagos Islands has been defended against the US government, which established an interest initially in the guano deposits for fertiliser, and latterly for geostrategic reasons during the Second World War. In 1944, taking advantage of Ecuador's recent defeat by Peru, the US government proposed a ninety-nine-year lease on the islands in return for economic assistance. Ecuador did not accept.

Box 2.3 The Peru–Ecuador territorial dispute

The contemporary dispute revolves around the Oriente (Amazon) province, and originates with the 1842 claim that followed the sale of land to generate funds for debt repayment. Peru was opposed to the selling of land which it considered its own. A Peruvian invasion of the territory followed at a time of Ecuadorean civil war in 1859. The Peruvians occupied Guayaquil and forced the signing of the 1860 Treaty of Mapasingue which cancelled the land sales and acknowledged Peruvian sovereignty. After the civil war, Ecuador expelled the Peruvians and failed to accept the validity of the treaty.

From 1887, Peru and Ecuador agreed to arbitrate the boundary issue. However, the importance of rubber within the region put the resulting 1890 García–Herrera arbitration treaty under pressure since Ecuador held the Amazon region. The Peruvians failed to ratify the treaty and both countries failed to accept a Spanish compromise agreement in 1910. Only inter-ventions from Argentina, Brazil and the USA prevented war. The USA attempted to arbitrate in 1924 and 1938 but Peru's *de facto* occupation left little opportunity for diplomacy. War between the two countries broke out in 1941, leaving Ecuador with half of its national territory in 1942, having lost the Oriente province. The Rio Protocol settling the territorial division at the close of the war was signed by Ecuador under US pressure. The protocol was denounced by Velasco Ibarra in 1960 and confrontations in 1978 and 1981 resulted from Ecuadorean demands for renegotiation of the territorial divisions in 1976, and for the protocol to be reviewed within the context of the 1978 Amazon Basin Treaty and the Andean Pact.

The outbreak of hostilities in the Oriente province in 1994 revealed the ongoing failure to reach an agreed settlement and the redundancy in late twentieth-century disputes of the territorial divisions at independence. Ecuadorean public documents carry the slogan: 'Ecuador always has been, is, and will be an Amazon country.' This is despite the fact that the Amazon region is principally undeveloped and inhabited by indigenous groups weakly connected with the nation-state. The central importance for the Ecuadorean and Peruvian governments lies in resource potential, such as oil reserves, hence both countries' attempts at integrating their eastern regions into the national whole and promoting the defence of these regions, although this falls short of administrative and economic decentralisation. This is in marked contrast to the significance in the post-independence years when the advantages of river basin links to the Atlantic were considered vital.

Sources: Castillo (1986), Corkill and Cubitt (1988), Bolívar Castillo (1993), Viteri Durand (1994), St John (1994)

Propaganda similar to the Ecuadorean slogan that it is 'an Amazon country' can be found throughout the continent. Chile lays claim to large tracts of the Antarctic while Argentina advertises its sovereignty over the Malvinas at its border entry points. Arbitration of claims and counter-claims will be resolved only on a case-by-case basis and will require agreement and the preservation of continued diplomatic ties to ensure that the treaties and settlements are adhered to. Nevertheless, as long as national sentiment and military preoccupations revolve around territorial integrity and the pursuit of historical claims, disputes and their unstabilising influences will continue. For Ecuador the outcome of all its territorial disputes is that it now controls only a third of its 1830 territorial space. Not until the territorial public cultures that are constructed at the national level for public educational purposes are renegotiated and reconstructed can the likelihood of boundary conflicts be reduced. Control over the development of geography curricula for schools and higher education institutions will play an important role if there is to be a change in the perpetuation of fierce national territorialism (Escolar *et al.*, 1993; Radcliffe and Westwood, 1996).

The Ecuadorean and Paraguayan cases reveal the complexities of the historical precedents embodied within the numerous contemporary geopolitical disputes within Latin America. While many disputes lie dormant for decades, resource finds, migration and periods of nationalist fervour lead to the re-emergence of diplomatic claims and counter-claims. Few of these occasions lead to armed conflict, but the recognition that boundary disputes provide the points of antagonism between Latin American nation-states dating back to the independence period leads to a better understanding of how populations have been educated into nationalist perspectives on their territorial spaces and spaces lying beyond.

As long as Latin America's military forces pursue territorial objectives, often based on early twentieth-century geopolitical theory, and can mobilise the national populations around these objectives, boundary disputes will continue to play a leading role in international relations within the continent. This will happen in spite of the apparent predominance of economic integration and neoliberal policies as the basis for inter-state linkages.

FURTHER READING

Edwin Williamson's *The Penguin History of Latin America* (1992) is an excellent detailed history of the pre-Columbian period to the late twentieth century. Burkholder and Johnson's *Colonial Latin America* (1994), Lockhart and Schwartz's *Early Latin America: A History of Colonial Spanish America and Brazil* (1983), and the Steins' *The Colonial Heritage of Latin America* (1970) all provide a broad coverage of the Iberian colonial period. Jack Child provides the best overview of contemporary geopolitics in *Geopolitics and Conflict in South America* (1985) with further material in the volume edited with P. Kelly, *Geopolitics of the Southern Cone and Antarctica* (1988). Other case studies of contemporary geopolitics can be found in Pascal Girot's (ed.) *World Boundaries*, vol. 4, *The Americas* (1994) and Lawrence Herzog's (ed.) *Changing Boundaries in the Americas* (1992).

3

LATIN AMERICAN GEOPOLITICAL ECONOMY

Latin America ... finds itself today in a sadly paradoxical bind. The end of the Cold War has brought greatly broadened geopolitical leeway, but economic globalization and ideological uniformity have rendered that at least partially meaningless.

(Castañeda 1994, 48)

THE NEW WORLD (DIS)ORDER AND GLOBALISATION

A New World Order?

The journalist Walter Lippman popularised the term 'Cold War' in 1947. It suggested a political, economic, propaganda and intelligence war in contrast to a 'Hot War' characterised by direct military engagement. He used the term with reference to the ideological conflict between the USA and USSR backed by their international support forces. While the Cold War ensued between the two Northern 'superpowers' from the late 1940s to the late 1980s, Latin America became incorporated into the stand-off as a theatre of 'hot wars', so-called 'surrogate' or low intensity conflicts, and ideologically inspired political and economic competition.

For most of the late twentieth century, the Cold War has provided the framework for Latin American geopolitics and geoeconomics. Only since the early 1990s has the geopolitical script changed with the arrival of the so-called New World (Dis)Order. For many, this post-Cold War reordering suggests new opportunities for polities and economies. For others, such as Noam Chomsky (1993), a 'new imperial age' of predominant supranational and transnational financial and commercial organisations has replaced the ideological militarism of the Cold War.

The re-emergence of political geography in the late 1970s and early 1980s coincided with a time of apparent geopolitical stability. The bipolar world of the Cold War predominated and the ideological confrontation promoted geopolitical thinking along the lines of statism and militarism. This MAD (mutually assured destruction) stability persisted until 1989, the global order remaining firmly fixed by the nuclear threat. Within this order, states peripheral to the superpowers were suspended, whether aligned or non-aligned, within a global economic and military balance. Such is the commonly held view, but it is difficult to equate this period of global geopolitical equilibrium with the succession of events within South and Central America. With considerable rimland activity and domino objectives, peripheral 'hot' wars sponsored by the superpowers ensured the self-satisfying stability of the Cold War belligerents.[1]

For Central America, as for many other Southern regions, the Cold War was far from a stable period since the region became an ideological battleground. In this regard it is interesting to note the degree of stability and peaceful negotiation now accorded Central America *since* the geopolitical transition was set in motion in the region, following the collapse of the Communist bloc, the fall of the Berlin Wall, and the fiasco of the Panama invasion in 1989. The Cold War balance of power has been replaced by a more complex set of power relations at inter- and intra-state levels of analysis.

The end of the Cold War has created a rapidity of change unprecedented since the end of the Second World War. The world (dis)order is under review yet does not fit the template of state control and organisation that political geography has always worked from. The state of flux that is being experienced offers little security, rather great uncertainty.[2] Within this geopolitical transition between world orders (Taylor, 1993b), the reorganisa-tion of linkages at many different levels, the issues of ethnic-territoriality such as nationalism and tribalism, the environment and resource manage-ment, social movements, fundamentalism and market and technological innovations, have all come together in a kaleidoscope of complex develop-ments and interactions.

The complexity of intra-state demands and organisation for change has put a new face on many Southern regions. The New World (Dis)Order that has emerged from the ending of the bipolar confrontation is multi-faceted and demands new theoretical perspectives. On an inter-state level, activities are increasingly complicated by transboundary agencies and processes. Most sovereign entities remain intact, but the forces of fragmentation are eroding states from within and without as international movements and international actors challenge the control and authority of individual states. While this erosion does not suggest that states are being hollowed out to any large degree leading to state ineffectiveness, it is true to say that there are new pressures on nation-states that require state flexibility and new approaches to organisation and control.

Until this point I have referred to the New World (Dis)Order, by which I mean a simultaneous ordering and disordering of the world, two antagonis-tic sets of processes acting at global and local scales. The debates surrounding the existence or non-existence of a New World Order are open to endless speculation due to the numerous theoretical standpoints taken in order to explain events and circumstances. As Adam Roberts (1991, 509) describes it, 'familiar compass-points have disappeared'. To suggest that the world order has undergone a geopolitical transition since the end of the 1980s is generally agreed upon, even if one takes the less supported view that this transition is one of sustained Cold War and that the third period of *détente* has just recently passed (de Blij, 1992). The complexity of this semantic debate reflects the problems of establishing a perspective on 'the global' in the late twentieth century.

Central America and the post-Cold War period

For Central America, the transition from the 'lost decade' of the 1980s that marked the final decade of Cold War to the process of pacification of the 1990s has been critically important for peace and development within the continent.[3] As a transition between the superpower conflict and the next world order, whatever that might be constructed as, there is currently an 'ideological interregnum' within the region (Thrift 1992, 5). The response of the region's governments has been to move towards greater autonomy in both political and economic spheres within this ideological power vacuum.

Within the movement towards a multipolar core at a global level to replace the bipolarity of the Cold War (Buzan, 1991), Central American governments believe themselves to have a greater opportunity to distance themselves from US political and economic pressure via links with the other global cores such as Japan and the European Union. These options will reduce the influence of the power relationship wielded by the USA, which has existed since the proclamation of the Monroe Doctrine and which was so evident during the Cold War, terminating in the Panama invasion.

The obstacles facing the Central American republics in any reformulation of global order are little different from those existing previously. US influence in the region since independence in the early nineteenth century, when the region actually existed as a territorial entity, has changed little although strategies have changed. Jorge Castañeda (1993a, 94) describes US–Latin American relations as a theatre of the absurd because of the wide range of strategies that have been employed. The persistence of the Monroe Doctrine within US foreign policy has been relentless in that economic stability ensuring unhampered transfer of Central American commodities has been protected regardless of regime type, usually in the name of national security. Robert Holden (1993), in a detailed assessment of US military involvement in the region, notes that low intensity conflict has been the mainstay of what he terms 'the real diplomacy of violence'.

Of significant interest, especially in Latin America, is the manner in which this low intensity conflict of the 1980s has been adapted to the 1990s with the concept of low intensity democracy. Barry Gills and Joel Rocamora (1992) regard 'democracy' as the 1990s buzzword to replace 'development' and in this sense the USA has altered its stance relative to Latin America, although the relationship between low intensity democracy and the promotion of the neoliberal model must be examined. To what extent the encouragement given to Central American governments to establish and maintain a democratic apparatus is yet another form of US involvement is open to speculation.[4] Persistence through pragmatism, with defined objectives, has maintained a considerable degree of distrust within the isthmus, even among allies such as Violeta Chamorro after the UNO coalition victory in Nicaragua in 1990 and Guillermo Endara following the Panama invasion of 1989.

During the Cold War, the region functioned within the US geostrategic realm dictated by ideological orientation (Cohen, 1973, 1982). It remains to be seen whether the region stays so firmly entrenched in the emergent US geomarket realm, although the extent of 'Coca-Colarisation', and cultural homogenisation via consumer goods, to date suggests that the region will be hard-pressed to remove itself from US corporate influence.[5] For US foreign policy, the NIMBY (not in my back yard) scenario remains, but the shift from political ideology to cultural idealism should be accounted for by political geographers. Of those who legitimise foreign policy (the experts, foreign policy professionals and the media), it is the media that leads in this recognition (Dodds, 1993a).

The predominant policy of US economic protection under whatever guise, maintaining necessary primary commodity imports and levels of consumer goods exports, is no less important post-Cold War than previously. The impact of NAFTA upon this US–Central American relationship has yet to be seen, but if Mexico is able, or willing, to provide more of such primary commodities and absorb more US manufactured goods, the relationship may well change considerably. If, as Miguel de Oliver (1993) suggests, NAFTA represents declining US hegemony, greater autonomy for the region must surely follow. Others regard a triangular commercial network (Central America–Mexico–USA) as likely to provide greater economic benefits for the isthmus (Gorostiaga, 1990).

A factor that may ensure that the USA keeps a careful distance in its dealings with Central American states is the threat of being unable to exert its dominance over a united regional front. Andrew Hurrell (1992) refers to this disunited situation as 'alphabet soup regionalism', whereby different bodies seek to assume responsibility for various areas of policy, as in the European case. By maintaining economic authority within the region, the USA will continue to manipulate Central American political economies.

The shadow of the US invasion of Panama in 1989 disturbs those who forecast greater autonomy for the region. As the first invasion of Latin American territory since the Second World War that could not be put down to superpower ideological confrontation, the Panama invasion was symbolic. While one may speculate as to what promoted the US action, apart from the displacement of the 'renegade' General Noriega, even including the suggestion that it was to curb a disturbing increase in Japanese investment in Panama, its impact upon Central America as a whole was powerful. The governments of the region, while desiring continued economic aid, were also desirous of US *perestroika* for the region. The US response to Central American anxiety was not characterised by reticence, but rather more by involvement.

The globalising political economy

What will shape US engagements with Central and South America into the next millennium is the evolution of the global neoliberal political economy and the processes of globalisation that characterise it.[6] Within the New World (Dis)Order of the 1990s, the concept of globalisation has gained increased currency alongside processes of deterritorialisation, the flow of capital and technology across national boundaries that reduces the importance of national space in economic decision-making (Corbridge, 1994).

Although it is derided by many for its universalising pretensions, the concept of 'the global' is an appropriate one for the end of the twentieth century. The global transcends the international, which maps the world around the reference points of the inter-nation-state system. The global seeks to introduce other transnational players and processes to the map, such as companies, people (as workers, migrants and refugees), and also the impacts of new technologies such as electronic and digital communication. These influences within the world work within and between nation-states, somewhat complicating traditional features of the nation-state such as territoriality and sovereignty. For example, a comparison of the capital accounts of particular nation-states and large multinational firms confuses the picture of global financial power and economic influence: Agnew and Corbridge (1995, 170) note that the market value of the Royal Dutch/Shell Group (Netherlands/UK) is US$77.82 billion while the GDP of Bolivia stands at only US$5.02 billion.

In a global political economy, firms differ financially from nation-states. In many cases the difference may only be the lower debt that they are saddled with. An analysis of international relations that looks only at the inter-state system and its geopolitics turns a blind eye to the geoeconomic and the geosocial forces operating upon the world map. The idea of the global suggests that these forces can be represented on the world map.[7]

Osvaldo Sunkel (1995, 46) identifies the globalising influences of the 1990s as fundamental historical changes:

> we are in the midst of a period of epochal civilizational transformation, a sea change in the sociocultural affairs and arrangements of humanity, within and among societies and their environmental life-support systems.

The influence of globalisation on the organisation of Latin American space

Plate 3.1 The geoeconomics of business and trade has transformed particular regions of Latin America. The port of Valparaíso (Chile) has been at the heart of Chilean commercial development since the early nineteenth century and acts as a maritime artery for Santiago's Metropolitan Region.

Plate 3.2 The southern island of Chiloé, Chile's poorest province, contrasts with Valparaíso and the Metropolitan Region. The geoeconomics of the island are still based on small-scale fisheries (exemplified by the proximity of housing to waterways and the sea) and other artisanal activities, although commercial fish farming is rapidly changing this situation.

is critically important to an understanding of late twentieth-century political geographies in the region. Changing emphasis on types of production and the location of this production, and the labour components of these processes, leads to the utilisation of new productive spaces and the dereliction of older activities and their surroundings which are no longer viable within the global marketplace. While the term globalisation seeks to explain a complex net of dispersed activities throughout the world, the impact on national and local spaces is far from abstract. Gereffi and Hempel (1996, 18) note that 'Economic globalisation is not a frictionless web of arm's-length market transactions.' Impacts of globalisation are direct and experienced by all members of society, often leading to social conflict as a result of the tensions within the processes of globalisation (Mittelman, 1994, 440–441):

> The mosaic of globalisation reflects a transformation of poverty in which three continents were most adversely affected by globalisation to the marginalisation primarily of a single world region and of enclaves in other regions ... Against a backdrop of transformation from a hegemonic and state-centred structure to a multipolar and politically decentred world system, globalisation is both an agent and product of social conflict.

The globalisation process in itself neither improves nor deteriorates the position of Latin American states, their economies and their workers and other citizens within the economic space of the global. Globalisation is similar to the internationalisation of production that preceded it during the post-Second World War period, the nature of commercial and trading links that became established from the post-independence period, the role of merchant houses within the nascent Latin American republics during the early nineteenth century, and the activities of the colonial period. Their common characteristics are those of the perpetuation of the exploitation of Latin America's resource base, its labour markets, and its low socio-economic condition relative to other regions of the world. While international finance is pumped into the region, the net value derived from capital, labour and resources that flow out of the region is more significant.

The nature of the evolution of the international financial system has lent itself well to globalisation. As a concept, globalisation is characterised, to a great extent, by the process of access to and integration within (engagement with) information flows, and also time-space compression.[8] As demands for capital have ceased to be confined to sources from within the nation-state of origin of the firm, capitalisation has taken firms outside many of the regulatory boundaries set by nation-states. The Bretton Woods Agreement (1944) and the founding of the International Monetary Fund, the World Bank and the General Agreement on Tariffs and Trade in the 1940s established and swiftly extended this idea of the internationalisation of capital.[9]

Nigel Thrift and Andrew Leyshon (1994) suggest that Bretton Woods marked the shift from the nineteenth-century nation-state-based financial system to one run by 'money capitalists'; this money power led to the creation of the so-called 'phantom state', a weakened nation-state structure. The super capital highways of financial transfers and dealings of the 1980s and 1990s, 'casino cyberspace', or the 'hypermobility of capital' (Cox 1992, 427) have mitigated against effective inter-state controlling mechanisms. The Latin American debt crisis from the early 1980s underlines this point. It is within this capitalist framework, run by neoliberal economic programmes, that Latin American nation-states operate and are operated by their socio-economies.

POWER, SPACE AND ECONOMIC DEVELOPMENT

A historical context

Contemporary Latin American geoeconomics, the power-space relations of production and consumption, has to be understood in terms of its historical development. From the pre-Columbian societies' trade in precious metals and natural resource products, to the *conquistadores'* search for El Dorado and their exploitation of silver at Potosí, Latin America's economic power has been established from its natural resource base: minerals, forestry products, agricultural and livestock products, and fisheries. While the twentieth century has seen a drive towards manufactured production, Latin American economies still rely on their traditional economic base (Tironi, 1985). Approximately 80 per cent of Latin America's exports are still natural resource-based.

This situation has led to two fundamental obstacles relating to the continent and its productive spaces. The first is that the ongoing relative decline of commodity prices relative to manufactures continues to plague balances of trade within the continent (Rowling, 1987). Only with higher commodity prices in international markets, or an increase in manufactures in the export equation or in the domestic market, can this trend be altered. The second is the state of the environment and the pursuit of sustainability. Once natural resources are exploited or extraction costs rise beyond competitive levels, many Latin American economies will lose vital revenue sources emanating from export-derived foreign exchange.

These two obstacles date back to independence. From the early eighteenth century, US and European merchant houses were active within the continent, establishing themselves as the pivots of national import-export trades, and as credit sources and national economic advisors (Greenhill, 1977; Albert, 1985; Miller, 1993). Their influence within the new republics in terms of establishing new trade patterns delinking the continent from its Iberian ties was critically important. Their manipulation of key markets, such as guano in Peru, cocoa in Ecuador, livestock and cereals in Argentina and Uruguay, and nitrates in Chile, ensured their predominant position within economic circles and élite society. An example of the power of the merchant houses was the US merchant house Grace's cancellation of the Peruvian state's debt in exchange for the lease for control over the railway system, a 4.5 million-acre Amazon forest grant, 3 million tons of guano, and thirty-three annual payments of $30,000 (Reid, 1984).

It was not until the early twentieth century that autonomous development strategies put the national economy back into the hands of the Latin American state. This came about as a response to the events of the First World War and the gradual undermining of the traditional merchant house commercial sector due to the 'multinationalisation' of US and European firms (Wilkins, 1970, 1974; Albert, 1983). Latin American states, if not still entrenched in specialised primary export commodities, engaged in alternative development programmes, linking economic modernisation via industrialisation with social programmes. Osvaldo Sunkel (1995) identifies the areas of state involvement: intervention in goods and factor markets; financing of long-term projects and programmes; redistribution of income; public investments in infrastructure; creation of public enterprises; and the generation of employment opportunities.

In terms of the organisation of national territory, these industrialisation policies did little to alter existing land ownership patterns and rural economic organisation. Only in urban areas did extensive changes result. Increased capitalisation of agribusiness and the emergence of a Latin American urban proletariat and industrial labour demand led to a transformation of urban spaces. Numerous bureaucratic-commercial entrepôts and colonial administrative centres were reorganised into urban-industrial centres during the late nineteenth and early twentieth century (Scobie, 1972). After the Second World War this process was accelerated with the internationalisation of multinational production in the 1960s and 1970s, leading to 'golden' periods of industrialisation such as Brazil's 'economic miracle'. Alongside this economic transformation, the continent became increasingly urbanised in terms of its demographic dynamics. This industrialisation process was typified by the role of the automobile industries, particularly in Mexico and Brazil. From a period of rapid growth in this sector from the 1950s, the 1980s revealed the underlying structural problems associated with the role of multinational corporations (MNCs), despite the Brazilian policies of a high domestic content requirement for automobiles produced in Brazil by MNCs (Shapiro, 1996).[10]

The shift in focus to the industrial in preference to the agricultural did little to antagonise the landowning élites since it was their capital, generated from semi-capitalist-semi-feudal economic operations, alongside foreign direct investment, that was transforming urban regions. Absentee landowners invested in domestic manufacturing to maintain the traditional power structures of Latin American societies. As in the immediate post-independence period, domestic élites became bound up with the foreign commercial and productive élites, creating a multinational bourgeoisie. The expanding urban labour force provided cheap inputs into the industrial programmes, establishing a compliant urban proletariat.

Within a fifty-year period during the middle decades of the twentieth century, Latin American rural farmers and peasants had been converted from the rural poor to the urban poor as a result of changing land use and new modes of production. In the same way as the rural poor have little recourse to social justice within the semi-feudal-semi-capitalist social and economic hierarchy, there are few cases of urban labour unions succeeding in opposition to monopoly capital. One of the few successes was the publicised

Table 3.1 Population, output and per capita GDP, 1975–94

	Population (millions)		Growth rate of population (% annual change)		GDP per capita (1988 dollars)		% growth of real GDP per capita (1988 dollars)	
	1984	1994	1975–84	1985–94	1984	1994	1975–84	1985–94
Developed economies	769	818	0.6	0.6	15,973	19,429	2.1	2.0
Economies in transition[a]	371	391	0.8	0.5	5,286	3,448	n/a	–4.2
Developing economies[b]	3615	4411	2.1	2.0	768	939	1.7	2.0
Latin America	390	474	2.3	2.0	1,975	2,079	0.7	0.5

[a] Eastern Europe and the former Soviet Union.
[b] 93 countries that account for 99% of the population of all developing countries.
Source: UN World Economic and Social Survey, 1995

Table 3.2 Latin America: gross domestic product

	Annual growth rates					Cumulative variation in GDP per capita			
	1991	1992	1993	1994	1995[a]	1981–90		1991–95[a]	
Latin America	3.6	3.1	3.3	4.5	1.7	12.7	–8.3	17.2	6.9
South American countries[b]	3.6	3.0	4.2	5.0	3.5	10.9	–9.2	20.8	10.7
Mexico, Central American & Caribbean countries[c]	3.5	3.2	1.1	3.4	–2.9	17.1	–5.9	8.5	–2.1

[a] Preliminary estimates.
[b] Argentina, Bolivia, Brazil, Chile, Colombia, Ecuador, Paraguay, Peru, Uruguay, Venezuela.
[c] Costa Rica, Dominican Republic, El Salvador, Guatemala, Haiti, Honduras, Mexico, Nicaragua, Panama.
Source: ECLAC, Economic Panorama of Latin America, 1995

Table 3.3 Growth of per capita GDP (at constant market prices; average annual rates)

Country	1970–80	1980–82	1982–85	1985–90	1993
Argentina	1.1	−5.9	−3.5	−1.1	4.8
Bolivia	1.4	−3.7	−3.8	0.2	1.7
Brazil	6.1	−3.6	−0.8	0.0	2.3
Chile	0.9	−5.3	−2.1	4.4	3.8
Colombia	3.2	−0.5	0.4	2.9	3.1
Costa Rica	2.6	−7.6	−2.7	1.5	3.6
Ecuador	5.8	0.0	−0.7	−0.7	0.0
El Salvador	0.7	−8.0	−2.9	0.0	2.4
Guatemala	2.8	−4.0	−3.9	0.0	0.8
Haiti	3.0	−4.8	−2.8	−1.8	—
Honduras	2.3	−2.6	−1.8	0.1	3.4
Mexico	3.6	1.6	−0.3	−0.6	−1.2
Nicaragua	−2.8	−0.7	−2.2	−5.8	−4.7
Panama	2.6	2.1	0.5	−3.4	3.6
Paraguay	5.6	0.5	−0.9	0.8	1.1
Peru	1.1	−0.2	−2.6	−4.1	4.4
Dominican Republic	4.2	0.3	−0.6	1.4	8.9
Uruguay	2.6	−4.5	−3.3	3.1	0.9
Venezuela	−1.6	−4.3	−4.5	0.3	−2.4
Total	3.1	−2.5	−1.5	−0.1	1.4[a]

[a] Excludes Haiti
Source: CEPAL, Statistical Yearbook for Latin America and the Caribbean, 1994

case of the Coca-Cola bottling plant dispute in Guatemala, resulting in plant sit-ins and occupations from 1984 to maintain a plant union, and eight workers' deaths in company reprisals. The labour dispute succeeded mainly as a result of international solidarity and pressure (Gatehouse and Reyes, 1987). International and domestic élite capital provides the source for these changes in social organisation and social relations. To analyse changing territorial organisation since independence is to analyse these changing relations of the Latin American state with international and domestic capital and interest groups.

Export-led development

The wave of export-led development that accompanied postwar foreign direct and portfolio investment in Latin America had a marked effect on the organisation of productive spaces. The promotion of suitable (non-traditional) products for the international market over more traditional products from the 1980s has had an effect on ownership, food security and the environment (Barham et al., 1992). Chile, hailed as the neoliberal success story since the mid-1970s, has witnessed marked changes in its economic geography. The drive towards newer 'dynamic' export sectors such as forestry, fruit, wine and salmon has led to the transformation of rural areas (Gwynne, 1993). Traditional forms of production are replaced by capital-intensive methods and new forms of ownership are introduced, replacing communal and smallholding practices. Costa Rica has followed a similar path, focusing on textiles.

Export-led development has led many countries to return to their natural resource bases in preference to industrialisation policies reminiscent of the failed ISI programmes. However, the under-valuing of environmental costs within the economic equations of national and international economists is having serious long-term effects (Carrier, 1991; Instituto de Ecología Política, 1994). It is only recently that economists working on Latin America have confronted the environmental costs of sustainable development and environmental damage in terms of underlying economic costs for the long term (see Meller *et al.*, 1996; Chisari *et al.*, 1996). Without sustainable development strategies for exploitation of the natural resource base, the comparative advantages of Latin American territorial spaces will be consumed. International capital will then move on. Chile and Costa Rica, two models of the neoliberal experiment, both have high dependency on natural resources and value-added products derived from them. There is little evidence to suggest that these neoliberal strategies are sustainable ones since opportunities to generate more sophisticated production processes, in order to sustain export growth with higher value-added content, appear unlikely to be introduced in the near future (Paus, 1994).

Many of Latin America's economic problems are founded upon the inability to provide more sophisticated products for the global marketplace, which was the ISI and intra-regional economic organisation dilemma. While East and South-east Asian nations drove towards high value-added products and established niches in the global marketplace with strong state policy backing from the 1970s, Latin American economies have been unable to do likewise. The Latin American continent is marked by the nation-state divisions between the producers of manufactures and the producers of raw commodities and their derivatives. In 1992, Brazil, Mexico, Argentina, Colombia and Venezuela contributed 90 per cent of Latin American manufactured exports, while Brazil and Mexico produced 85 per cent of the region's most advanced industrial products such as computers and electronic goods (Gereffi and Hempel, 1996).

With the economic trends during the Cold War period passing from corporatism as characterised by Fordist and Taylorist mass-production methods, often with high degrees of state intervention, to flexible specialisation characterised by 'just-in-time' and other methods of rationalisation within smaller-scale business units, firms have multiplied and sought capitalisation from the more diverse sources of credit and investment now available within the global financial markets. While this process has been

Table 3.4 Structure of trade in goods and services (selected countries)

	Manufactures			Non-fuel primary commodities			Travel receipts and remittances		
	1980	1985	1992	1980	1985	1992	1980	1985	1992
Latin America	13.3	20.1	27.2	31.7	30.3	28.9	6.3	7.0	11.0
Argentina	16.6	17.4	20.1	52.5	58.0	49.5	3.1	5.1	4.9
Brazil	33.4	39.2	50.6	51.6	42.7	35.5	0.6	0.2	2.4
Mexico	8.4	22.4	32.0	14.9	10.6	10.8	17.7	13.9	19.7
Peru	11.4	8.7	16.4	42.2	46.6	57.9	6.1	7.6	4.2

Source: UN World Economic and Social Survey, 1995

primarily a Northern phenomenon, the same practices are increasingly being adopted on a global level. The export processing zones (EPZs) in El Salvador provide suitable examples.

EPZs, including the *maquiladora* sector, have generally been characterised by the development of textile and clothing industries in Latin America. Generally known as 'sweatshops', companies under contract from larger manufacturers supply the US market (principally) with garments assembled in the EPZs, taking advantage of tax exemptions, special exchange schemes, free utilities and complete profit repatriation where necessary (Figueroa, 1996). The reverse side of this economic growth trend in Latin America (particularly developed in Mexico, Central America and Caribbean states such as the Dominican Republic) is the poor labour legislation and regulation of this sector where low wages, child labour and a lack of job security leads to bad practices.

Large-scale firms are still critically important in terms of employment and production within the geoeconomic arena. However the global economically active population is overwhelmingly employed in small-scale businesses, both formal and informal.[11] The impact of the informality of workers is significant for nation-states in that such self-help strategies mean that many people are geoeconomically active yet contribute little to nation-state geopolitical health such as GDP. As such, geoeconomics has to take into consideration the social factors of labour, such as issues of the reproduction of the labour force, health and education.

The dynamic postwar growth period in Latin America ended in the early 1980s as the debt crisis and international recession affected authoritarian and democratic regimes equally. Duncan Green (1996) identifies three phases of

Table 3.5 Commodity composition of Latin American trade, 1980–92

	Imports			Exports		
	1980	1985	1992	1980	1985	1992
Primary commodities						
Food	6.0	4.8	4.4	14.2	15.8	10.9
Agricultural raw materials	3.1	3.1	3.4	4.6	4.0	4.6
Fuels	6.7	5.5	4.1	9.5	11.0	11.5
Ores and metals	3.1	3.8	2.8	10.9	13.3	13.5
Manufactures						
Textiles	3.6	2.4	3.2	2.2	2.3	1.7
Chemicals	8.0	6.6	5.3	2.0	2.9	2.3
Machinery and transport	7.8	5.2	4.9	1.0	1.5	1.2
Metal	6.4	3.5	4.3	1.6	4.2	4.3
Other	4.9	3.6	3.6	1.7	2.4	1.8
Total (US$ bill)	126.1	92.6	150.0	107.8	109.2	136.6

Source: UN World Economic and Social Survey, 1995

Latin American economic performance: 1982–83 – deep recession caused by the immediate response to the debt crisis; 1984–87 – a false dawn led by Brazil and Argentine 'heterodox' stabilisation programmes based on new currencies, exchange rate policies, and wage and price freezes; 1988–95 – stagnation while structural adjustment gathers pace, followed by patchy recovery.

Since the 1980s, the straitjacket of structural adjustment policies (SAPs) administered by the multilateral lending institutions (specifically the IMF and World Bank) has converted the state apparatus into a compliant, under-resourced entity subsumed within the flows and networks of globalised capital and private sector economic organisation. SAPs differ widely in Latin America due to the variations in national situations. The bases of the policies are those of economic indicator improvements. This is most often achieved by currency devaluations that have led to accelerating inflation and increased costs of living in many regions. Another measure that improves the indicator of public spending is that government subsidy programmes are cut. These subsidies often provide the life-line for the most vulnerable sectors of society since they may include basic foodstuff provision or subsidised credit to farmers (Safa, 1995).

Groups most affected by further cuts in social spending on education, health, housing and infrastructural improvement schemes can easily be defined: the poor, particularly women (who bear the burden of the shift of social responsibilities to the household), the aged and the young (who are dependent and suffer most health problems).[12] The high levels of social opposition to these austerity programmes are evidence of the despair among the poorest sectors of society. Food riots, a response to the failure of these programmes to meet basic needs, and strikes, a response to increasing formal unemployment and lower wages, are the principal manifestations of this opposition (Green, 1995). The riots are commonly known as 'IMF riots', highlighting the role of International Monetary Fund (IMF) policies which lie at the heart of the austerity measures.

The ability of the state to organise and develop national territorial space has been heavily diluted by the privatisation of large tracts of national space as a response to external economic pressures. Where once the domestic landowning élites dominated the organisation of space around agribusiness, international capital now adds its weight to these pre-existing structures. This situation builds on the early developments in the privatisation of Latin American spaces from the nineteenth century as land grants and economic opportunities attracted US and European capital and immigrants. The SAPs are little different from the merchant house commercial policies advanced during the early nineteenth century. They have both served to enable the continued flow of resources away from Latin America, principally natural resources and labour resources, to the Northern hemisphere.

REGIONAL PACTS AND INTERCONTINENTAL COOPERATION

The regionalisation of Latin America

One method of overcoming the neocolonial ties of extra-continental linkages that have characterised the continent's economic development and geopolitical orientation has been that of regional organisation and the establishment of trading blocs. By developing intra-continental relations, attempts have been made to counter the overbearing influence of relations with the USA and other core regions of the global political economy.

In terms of physical geography, Latin America forms distinct regions: the Darien Gap divides Central from South America; the Andean *cordillera*, East from West; the Amazon, North-east from South-east; the Atacama, North-west from South-west. Such divisions have played a critical part in the economic, social and cultural evolution of the continent. It must come as no surprise that the few large remaining indigenous Amerindian groups find themselves in physically remote, often impenetrable, areas such as the *altiplano* and the Amazon. Elsewhere, they have been integrated into, or have disappeared from, the socio-cultural landscape as a result of newly introduced economic activities and political-administrative territorial divisions.

From the pre-Columbian period of regional hegemonic societies such as the Aztecs and Incas, territories have been absorbed and reorganised, constructed, deconstructed and reconstructed. Regional agreements and disagreements were developed during the pre-Columbian period as Amerindian groups established territorial control and organisation of local spaces. Through mechanisms of conflict and arbitration, regional spaces were established and socio-cultural groups embedded therein. In the cases of the Aztecs, Mayas and Incas, imperial expansionary zeal led to territorial acquisition through military means. Despite these processes of regional integration, it was not until the post-colonial period that regional organisation in terms of a unifying strategy based on independent rather than imperial or colonial objectives came into being.

Gran Colombia and the Confederation of Central American States in the early nineteenth century provide examples of this unifying strategy. The latter's fragmentation in 1838 marked the end of a formal consociation of nation-states in Latin America. The nineteenth and early twentieth centuries were marked by considerable external orientation of the fledgling nation-states rather than inward orientation, since economic strategies were bound up with the USA and European commerce, trading primary commodities for manufactured and semi-manufactured products. In certain cases this commerce was to such a degree that the term 'informal empire' has been applied, although the degree of control of formal versus informal empires is difficult to establish (Thompson, 1992). Once again, neocolonialism provides a better term due to the absence of 'direct' political control in the majority of cases.

The First World War served to alter the international linkages of the Latin American states as US commercial and political influence extended southwards from its heartland in Mexico and Central America. These linkages were consolidated during the interwar period as Latin American economies suffered from protectionism in their principal commercial partner nations and from financial crises emanating from the 1929 financial 'Crash' and the ensuing Depression (Ground, 1988; Kindleberger, 1986). The Second World War served to hamper European–Latin American trade, as had the First.

The USA emerged in 1945 as the focal point for trade with Central and South America. Loans and other forms of military and economic assistance, often in order to support US business located within the region, combined with the strengthening of a Panamerican consciousness within the OAS, all worked in favour of the USA's role within the region. A notable example of the outcome of US influence in the region was the support of the US fruit companies in Central America. The CIA-backed coup in Guatemala in 1954 was an act in support of the United Fruit Company which had been subject to land reform policies. The US-backed administration that followed the coup terminated the land reform measures.

Only from the late 1950s did the Latin American republics attempt to orientate themselves more internally within the continent as a further step towards the promotion of Latin American industrialisation, marking an intended shift from their dependence on intercontinental trade. The ISI programmes developed from the interwar period in Latin America, as a response to the declining terms of trade between Latin America and US and European trading partners, had reached difficulties by the 1950s.[13] The inability to transcend the gap between consumer goods production and intermediate and capital goods production without high levels of imports had resulted in an impasse. The need for foreign capital and capital goods would only transform the dependent linkage, not reduce it (Macario, 1964; Díaz-Alejandro, 1965). Regional economic integration was understood to be a way out of the impasse since a sharing of technological developments from within the region and the construction of a considerably larger marketplace for products might enable firms to shift into the production of more sophisticated goods.

In 1960, the Latin American Free Trade Association (LAFTA) was established for South American countries to pursue regionally defined economic development patterns; Central American countries established the Central American Common Market (CACM). Having hidden their nascent manufacturing sectors behind tariff walls since ISI programmes were activated, the region's states slowly eliminated tariffs and restrictions on trade within their respective regions (Gwynne, 1994). Within a relatively short period of time, however, the structural obstacles present within the regions had become evident.

Problems were three-fold: individual governments had different expectations from the organisations and were often obstructive, continuing their pursuit of ISI and defending their industries; production within each regional bloc was too similar to enable useful flows of raw materials, semi-finished or finished products; and finally, the level of trade within the blocs remained low at 10 per cent of the total (Gwynne, 1994). The outcome of the failure of LAFTA was the formation of an Andean Group in 1969 (Chile [until 1976], Bolivia, Peru, Ecuador, Colombia, and Venezuela [from 1973]) and the renaming of the remaining group in 1980 as the Latin American Integration

Table 3.6 Intra-regional exports and imports[a]

Country	1970 Exp.	1970 Imp.	1980 Exp.	1980 Imp.	1990 Exp.	1990 Imp.	1993 Exp.	1993 Imp.
Argentina	21.0	20.9	23.6	19.0	26.0	26.6	41.4	31.9
Bolivia	9.6	19.2	35.7	51.4	44.8	48.2	37.3	66.5
Brazil	11.3	10.8	18.1	11.7	11.3	16.3	24.9	19.2
Chile	11.2	20.3	24.3	27.7	12.6	22.8	20.0	23.4
Colombia	13.3	8.4	16.6	19.8	16.1	18.4	25.2	23.6
Ecuador	11.1	13.0	19.2	14.4	17.7	22.0	21.0	23.0
Mexico	10.7	3.0	6.2	3.9	6.0	3.2	4.9	3.3[b]
Paraguay	38.5	39.5	45.7	59.2	52.4	39.2	49.0	54.2
Peru	6.5	15.5	21.2	15.0	14.6	33.1	18.6	37.7
Uruguay	12.6	29.3	37.3	37.4	39.5	48.1	51.2	57.2
Venezuela	12.5	3.7	14.1	9.5	8.9	12.5	20.0	16.2
Total Latin American Integration Association (ALADI)	12.7	11.1	16.4	13.4	13.0	13.2	18.7	16.7
Costa Rica	23.1	31.0	33.3	35.1	16.5	21.5	14.0	18.3
El Salvador	32.6	32.3	28.5	49.3	34.7	31.2	32.8	30.7
Guatemala	36.9	33.0	32.6	35.6	34.6	27.7	42.3	26.6
Honduras	16.7	32.3	13.5	32.0	6.5	18.4	3.8	16.6
Nicaragua	27.6	32.4	19.7	57.8	21.8	23.4	21.2	51.3
Total Central American Common Market (CACM)	28.5	32.2	27.4	40.6	23.1	24.7	20.7	24.6
Haiti	1.5	3.3	1.1	8.0	0.9	6.6	0.6	4.0
Panama[c]	4.9	13.0	19.2	11.1	18.2	10.2	19.8	6.4
Dominican Republic	0.9	5.6	10.4	22.7	3.2	11.8	3.2	16.2

[a] Percentages of intra-regional imports and exports with respect to total imports and exports, calculated on the basis of FOB (free on board) exports.
[b] 1993 includes *maquila* production.
[c] Excludes the Canal Zone.
Source: CEPAL, *Statistical Yearbook for Latin America and the Caribbean, 1994*

Table 3.7 Merchandise exports of MERCOSUR countries by region, 1990 and 1994
(US$ million)

Destination		MERCOSUR countries	Other Latin America	Total Latin America	All other regions	World
Origin						
Argentina	1990	1,833	1,577	3,410	8,943	12,353
	1992	2,327	1,900	4,227	8,008	12,235
	1994	—	—	—	—	15,659
Brazil	1990	1,320	2,397	3,717	27,697	31,414
	1992	4,099	4,045	8,144	27,649	35,793
	1994	5,921	4,706	10,627	32,931	43,558
Paraguay	1990	379	123	502	457	959
	1992	246	89	335	322	657
	1994	—	—	—	—	740
Uruguay	1990	595	94	689	1,004	1,693
	1992	544	135	679	1,024	1,703
	1994	896	142	1,038	875	1,913
MERCOSUR	1990	4,127	4,191	8,318	38,101	46,419
	1992	7,216	6,169	13,385	37,003	50,388
	1994	—	—	—	—	61,870

Source: CEPAL, *Statistical Yearbook for Latin America and the Caribbean, 1994*

Association (ALADI).[14] ALADI has less ambitious objectives than LAFTA although its intention is also to promote freer trade within the region. Since its inception, the Andean Group has suffered similar problems to those of LAFTA, aggravated by the lower purchasing power within the regional market and by political instability. Despite collective projects such as a regionally linked auto industry (similar to a LAFTA regional aircraft industry concept (Baerreson *et al.*, 1965)), intra-regional trade levels still remain low.

MERCOSUR (The Southern Cone Common Market), established in 1991 under the Treaty of Asunción and fully operational from 1995, provided a new face for the failed LAFTA set-up, and the undynamic ALADI for the non-Andean economies. As a regional grouping of the Southern Cone (Argentina, Brazil, Paraguay, Uruguay), MERCOSUR aimed for the construction of a free trade area. To date, intra-regional flows are markedly improved upon former levels, and the member states are continuing to reduce the trade obstacles. With Argentina and Brazil (and Chile entering in mid-1996) as members of the bloc, the internal market is large and the members relatively well-established players within the global marketplace. This should enable MERCOSUR to consolidate its position within the global arena, political divisions and financial crises withstanding.[15]

While South America has struggled to establish regional economic stability, Central America has demonstrated its ability to (re)construct a

Table 3.8 Merchandise imports of MERCOSUR countries by region, 1990 and 1994 (US$ million)

Destination		MERCOSUR countries	Other Latin America	Total Latin America	All other regions	World
Argentina	1990	1,833	1,577	3,410	8,943	12,353
	1992	2,327	1,900	4,227	8,008	12,235
	1994	—	—	—	—	15,659
Brazil	1990	1,320	2,397	3,717	27,697	31,414
	1992	4,099	4,045	8,144	27,649	35,793
	1994	5,921	4,706	10,627	32,931	43,558
Paraguay	1990	379	123	502	457	959
	1992	246	89	335	322	657
	1994	—	—	—	—	740
Uruguay	1990	595	94	689	1,004	1,693
	1992	544	135	679	1,024	1,703
	1994	896	142	1,038	875	1,913
MERCOSUR	1990	4,127	4,191	8,318	38,101	46,419
	1992	7,216	6,169	13,385	37,003	50,388
	1994	—	—	—	—	61,870

Source: WTO, *International Trade: Trends and Statistics* (1995)

regional identity since the turn of the 1990s, following the inability of CACM to promote any lasting region-wide economic development (although CACM has arguably been more successful than LAFTA). Since the early nineteenth century, when the Confederation of Central American states fragmented, the region has remained divided, competing for similar export markets, and experiencing successions of authoritarian political systems and US economic and military intervention. The end of the Cold War and of the 'lost decade' of the 1980s has, however, signalled a new development in terms of a Central American regional bloc. Momentum in this direction has gathered since 1989 as a response to reduced direct and indirect US state intervention and a realisation of the need to advance as a collective entity within a global ordering of similar groupings (Hurrell, 1992).

The Central American Common Market, a low key affair from its initiation in 1960, has been revitalised for regional economic interests alongside the Secretariat for Central American Integration (SIECA). On the political front the Contadora peace process and the two Esquipulas summits (1986, 1987) that were established to seek an end to the violence in the region initiated a degree of regional cooperation that has developed into a Central American Parliament and more effective inter-governmental relations (Dabene, 1991). This Parliament (PARLACEN) in many ways mirrors the European Parliament in that twenty parliamentarians from each nation sit as a regional forum. The first preparatory meeting was held in June 1991, consolidating the movement towards Central American integration. A further initiative came at the tenth meeting of Central American presidents in July 1991, when the Organisation of Central American states (ODECA) was reactivated and moves were made towards the region's insertion into the global market.

The most striking developments for regional coherence have arisen from Central America's increased links with the European Union through annual meetings since 1984. Via trade agreements, the provision of credits, and loans to revitalise the Central American Bank of Economic Integration (BCIE), the region has sought to diversify its ties.[16] However, the degree of European economic assistance has now reached a level whereby European pressures are becoming apparent due to the degree of capital and commodity linkages. Since the fall of the Berlin Wall, the attraction of Eastern European trade and investment opportunities has reduced capital flows to Latin America.[17]

It would appear that any Central American efforts to re-emerge from the 'lost decade' of the 1980s will be dependent on developments elsewhere, thus limiting the alternatives available (Davis, 1991). However, this is not to deny agency within Central America. The issue is clearly how particular regions are inserted, and insert themselves, into the global political economy. An obstacle for Central America, and similar Southern regions, is that cheap labour, new consumer markets and better investment returns are currently available within the northern hemisphere. These markets are in heavy competition with the select emerging markets in the 'South' that have been the destinations of considerable foreign direct investment in the early 1990s (that is, Chile, Mexico, Malaysia, Thailand). What is clear is that investment and finance dominated by the private sector now underpins US Panamericanism. The Monroe Doctrine still manifests itself in commercial strategies and diplomatic affairs. The outcome of this global condition of Northern private sector, principally US, influence is that a Panamericanist New World Order, dictated by the USA, may well prevent Central America from effecting a reorientation that distances itself from its northerly neighbour.

Regional blocs in the global political economy

Regional pacts have, on the whole, provided Latin American nation-states with the opportunities to redress the overbearing commercial ties that the majority of them had established with the USA. The failures of the 1960s have served as lessons but the regional pacts of the 1990s exist within a global political economy of regional blocs, thus facilitating a greater scope for new commercial networks than during the 1960s when the persistence of ISI proved to be a stumbling block. With neoliberal export-led markets characterising the regional groupings of the 1990s, alternative commercial opportunities are apparent. Robert Gwynne (1994) makes the following distinctions within Latin America in terms of contemporary regional integration: countries promoting unilateral integration with countries beyond Latin America: Mexico (entered NAFTA, 1994) and Chile (in discussions to enter NAFTA and entering MERCOSUR); and countries committed to regional integration within Latin America, within CACM (Guatemala, El Salvador, Honduras, Nicaragua, Costa Rica), the Andean Group (Venezuela, Colombia, Ecuador, Peru, Bolivia), and MERCOSUR (Argentina, Brazil, Paraguay, Uruguay).

The example of Chile, which is seeking entry into NAFTA and joined MERCOSUR in July 1996, and is also within APEC (Asia Pacific Economic Cooperation), points the way forward for the stronger economies that can engage freely in a multilateral context. However, the majority of Latin American economies are seeking the security of regional organisations in order to remain competitive in the face of global regionalism. The regional economic pacts, and, to a degree in the case of Central America, political alliances, have been integral to the construction of Latin American political and economic space during the second half of the twentieth century and will continue to be important during the 1990s.

While flaws in the attempts to establish internally free trading regions and integrated economic systems became apparent during the 1960s and 1970s, the return to new economic alliances in the 1990s has demonstrated the perceived need for economic integration within Central and South America in the face of similar processes and alliances elsewhere. By establishing freer economic boundaries between the republics for flows of capital and goods, the opportunities for greater intra-regional trade are improved. However, there will be a continued requirement for good political relations within the regions and a sense of collective identity in order for these regional reformulations to be more effective that their predecessors.

With the EU building tariff walls around itself and NAFTA attempting to construct an almost self-containing regional bloc, the ability of the GATT, thrashed out during the Uruguay round (1986–94), to maintain freedom throughout the global marketplace has to be questioned. The GATT is only as strong as its signatories, and their commitment to trade beyond their blocs in the future is debatable. The formation of regional blocs is a defence against the vagaries of the global political economy. The severe impact of the 1929 'Crash' and the 1982 debt crisis are lessons that have had to be learnt from. By integrating their economies on a regional basis, the Latin American republics are preparing themselves for the likely inter-bloc trade wars of the twenty-first century.

In terms of inter-bloc trade, the history of intercontinental ties follows on from the history of regional pacts and their internal coherence. The imperial heritage of the Latin American continent and, preceding the conquest, the movement of Asiatic peoples over the Bering Strait and into the Americas established the first intercontinental linkages in terms of social and economic links. During the post-independence period, these linkages have been based on commercial and financial ties, within neocolonial relations and, latterly, relations established by the financial insertion of the nation-states of the continent into the global capitalist economy. Cuba can be said to be the only republic to stand apart from these developments, since its own dependence was established via subsidisation from the USSR and trade within the Communist COMECON bloc. The most significant intercontinental linkages have been made by particular nation-states such as Mexico within NAFTA, or groups of nation-states within the regional organisations noted previously, such as CACM and MERCOSUR.

Arguably the key feature of commercial relations pertaining to the Americas in the late twentieth century has been the North American Free Trade Agreement (NAFTA), which came into being in 1994, building on the 1988 USA–Canada Free Trade Agreement. The agreement linked the USA, Canada and Mexico within an effective free trade area with key provisions relating to energy and resource trade, water commodification, public-private sector ownership, the reduction of tariffs and quotas, and a reduction in obstacles facing foreign firms (Campbell, 1991). From the outset it was clear that the intention of NAFTA was not necessarily to restrict itself to the Americas, or solely to North American states, since the agreement was not constructed as having spatial limits in terms of its evolution. The next member to board the NAFTA trade wagon will be Chile, although negotiations regarding the form of its entry, full membership or bilateral relations with the existing members have been delayed within the US Congress. As the much lauded neoliberal emerging market of Latin America, Chile has the suitable prerequisites to gain entry: an open, liberalised and deregulated economy with controlled inflation and high levels of foreign investment feeding into the privatising state sector.

While NAFTA is potentially the most significant contemporary regional bloc in terms of economic power and a large internal market, other forms of multilateral cooperation are also crossing over the various Associations' perimeters and influencing trade patterns. The Asia Pacific Economic Cooperation (APEC) conference is a new intercontinental forum in the global commercial arena. However, unlike the deeply integrated EU and NAFTA, APEC currently has no formal agreements and acts as a 'shallow-integration' strategy for the cooperating parties (Inoguchi, 1995). The conference consists of a large number of countries: Australia, New Zealand, the ASEAN nations, Vietnam, China, Taiwan, Hong Kong, Russia, Canada, the USA, Mexico and Chile.

Although it is still in its infancy, APEC is potentially the largest marketplace within the global political economy. It is still more interesting for two other reasons: the first is the breadth of political systems apparent among its members; the second is the potential for a shift of the centre of the global political economy away from the Atlantic for the first time since the rise of mercantilism. The Pacific Rim geographical reference represents a wide range of countries, despite the fact that this term is generally used for the Asiatic nation-states alone. The potential range of products, from raw materials and commodities on the Canadian, South and Central American western seaboards, to technology in Japan, within ASEAN, and along the US seaboard, and the size of potential markets, from the metropoli of California to Tokyo, Mexico City and Shanghai, suggest that APEC's development may well determine significant changes in the global political economy during the early twenty-first century.

In terms of the relations of Latin America as a whole with other regions of the world, it is important to note the emergence of Japanese business and investment within many Latin American countries (Stallings and Horisaka, 1994). Japanese trade and investment with Latin America increased dramatically at the beginning of the 1980s, in particular imports of beef and coffee from Brazil and Mexico and exports of technological goods, but both then faded again as a response to the debt crisis and international recession (CEPAL, 1988). However, from the late 1980s, Japan has been symptomatic of the rising commercial interest of East and South-east Asian business in the continent. For this reason, APEC provides an important cooperative starting point for the development of extensive commercial networks across the Pacific.

While European countries remain outside APEC, Latin America's extensive trade with European countries is still second only to its trade with the USA and Latin America negotiates on an economic bloc basis with the EU, as in the CACM–EU meetings that have taken place since 1984 (van Klaveren, 1994). Despite new investment flows into Latin America from elsewhere, there is still an important historical connection with European companies and European finance, dating back to the post-independence period (Roddick and O'Brien, 1982). As late as the 1970s, when the USA had established a commercial stronghold over the continent, 30 per cent of manufacturing was still European-owned within the largest South American economies (Jenkins, 1982). In an effort to establish a degree of multilateralism within inter-regional trading channels, the EU and its member states will continue to be important as trading partners of Latin American blocs and individual republics, despite transitions within the Pacific commercial sphere.

The significant difference in the contemporary structuring of these new commercial relations is the reduced level of geopolitical interference and a high degree of regional autonomy accorded to the regional blocs. Within the constraints of the neoliberal system and the strait-jacket of international financial pressures, which might be defined by some as a new imperialism manipulated by key Northern nation-states (Green, 1995), there are different opportunities for economic development within Latin America based on regional integration and inter-regional negotiations within the global capitalist economy.

Table 3.9 Latin America: direction of trade, 1980–92 (% of total)

	Imports from / Latin America	Exports to Latin America	Imports from / developed market economies[a]	Exports to developed market economies[a]	Imports from / economies in transition	Exports to economies in transition	Imports from / developing countries	Exports to developing countries
1980	18.2	21.3	60.5	64.4	4.1	6.5	35.4	27.5
1985	14.1	11.9	61.9	69.0	7.5	8.0	30.6	20.8
1990	13.8	13.6	72.1	62.8	6.2	4.9	21.7	21.4
1992	15.3	19.5	74.1	68.0	0.8	2.8	25.1	27.6

	Imports from / Africa	Exports to Africa	Imports from / West Asia	Exports to West Asia	Imports from / South and East Asia	Exports to South and East Asia	Imports from / Other Asia[b]	Exports to Other Asia[b]
1980	4.7	2.2	9.3	1.5	2.9	1.3	0.3	0.7
1985	2.5	2.5	9.9	1.9	3.2	2.8	0.7	1.5
1990	0.5	1.5	1.9	1.6	4.8	3.7	0.6	0.8
1992	0.6	1.3	1.6	1.7	7.0	4.1	0.6	0.8

[a] Including South Africa.
[b] Including China, Democratic People's Republic of Korea, Mongolia, Vietnam.
Source: United Nations, *World Economic and Social Survey, 1995*

THE DEBT CRISES

A cycle of crises

The two World Wars and the 1929 'Crash' and subsequent Depression marked important geoeconomic shifts within Latin America during the twentieth century, but it is the 'debt crisis' (the Second Great Depression) that arose during the early 1980s that has shaped the late twentieth-century political geography of the continent more than any other single set of events. The debt issue has led to a state of financial turmoil in the continent that William Guttman (1990, 161) describes as a 'permanent crisis' of repudiation, bankruptcy, default and systemic collapse. However, the repercussions of international debt cannot be understood purely through analyses of principal loans and interest payments. The impacts of the debt crisis have passed down to the most vulnerable sectors of society, who have initiated alternative survival patterns, such as different economic activities and new coping strategies. For treasury ministers and development economists, the debt is a numbers game. For the majority of Latin Americans it has meant significant changes to their livelihoods. Nora Lustig (1995, 61) stresses the social impacts of the debt crisis:

> [The 1982 debt crisis] has forced most countries to go through drastic adjustment and profound economic restructuring ... In most countries

the adjustment process has resulted in a drop in real average incomes, a reduction of subsidies for basic foodstuffs, and a reduction in public expenditure in the social sectors. It should be expected then that the crisis and subsequent adjustment have resulted in rising poverty and deteriorating social indicators.

The reasons for these changes are to be found in the policies that have been put in place to facilitate debt interest repayment, particularly those derived from SAPs, the austerity programmes that accompany international finance from the International Monetary Fund, World Bank and Inter-American Development Bank (see George and Sabelli, 1994). Highlighting the massive reductions in health and education spending sacrificed as part of the payment process, Jackie Roddick (1988) described the debt and the measures implemented for its servicing in Latin America as a social time bomb. The 'IMF riots' are a form of low intensity violence that supports such a pessimistic forecast.

One important point to bear in mind regarding the current Latin American debt crisis is that it is not unusual. Loans to Latin American republics and defaults on those loans have characterised Latin American finance since the early post-independence period (Marichal, 1989). Simón Bolívar sought a loan from the London capital market in the early 1820s, only to default on the repayments a few years later; other republics followed suit. After a period of unpopularity on the London financial markets, the same cycle of loans and defaults re-emerged in the 1860s, in the 1890s after the so-called Barings lending crisis in Argentina, again in the 1920s, and more recently in the 1980s.

In the same way that default disturbed the London capital market in the 1820s, the Mexican debt default in 1982 shook the financial institutions of the world to their roots. The key difference between the two cases was the scale: the former was concentrated on the London market, the centre of the international financial world at the time; the latter was global in scale and the exposure to the default was considerably more in relative terms. The Mexican default, in an instant, made the globalisation of finance apparent. The highly integrated international financial system that had injected OPEC petro-dollars via Europe and the USA into Latin America during the 1970s was immediately held under the microscope. The high sums involved in the newly acclaimed 'debt crisis' threatened both lenders, who had exposed themselves well beyond their capitalisation, and the debtors, who required the flow of new loans to maintain burgeoning public sectors.

The 1982 crisis

Latin America has been the focus of the 1980s debt crisis despite the fact that there is a clearer North–South dimension to the debt infrastructure and its associated problems. The reasons for Latin America's centrality is that the largest debtors in the global financial market are Mexico and Brazil. At the other end of the scale, the sum of the Bolivian debt is considerably smaller but ability to pay interest payments is considerably less. For this reason, Bolivian debt has been granted the 'Toronto treatment', normally reserved for sub-Saharan countries, of the condoning of a third of the bilateral debt, and the rearrangement of loan payments and, correspondingly, interest rates (Farthing and Villegas, 1991). Rather than the size of debts, the key considerations are ability to pay, at what interest rate and over what time-scale (see Table 3.10). These have been, and are, the debates within international financial circles, between commercial banks, government treasuries, the multilateral lending institutions (the IMF, World Bank), and 'independent' arbitrators, such as British Chancellor Lawson and US Secretary of State Brady during the 1980s.

Fifteen years down the road of the debt crisis, the lenders who did not immediately 'fail' have weathered the storm via loan loss provision, state intervention and debt renegotiation, while the debtors have been subjected to harsh externally imposed economic reforms. The outcomes of the debt crisis and the ensuing reforms have been diverse. For instance, migration to urban areas has continued unabated despite the urbanisation policy measures put in place; cities continue to 'spread', suburbanising and developing beyond the flexible peripheries of the urban system; and urban morphology is shaped by the effects of the external pressures on business.

Mexico City provides a good example of a city deteriorating under debt repayment, with grave socio-economic repercussions (Buchhofer and Guillermo Aguilar, 1992). Migratory patterns have resulted from agrarian change that has promoted new foreign exchange-earning goods for traditional goods with concomitant effects on product prices and land ownership. The SAPs have emaciated the public sector in urban centres, opened up internal markets at the expense of domestic entrepreneurialism, and increased the informal sector as a result.

Against this background of enforced neoliberal austerity, the impact of the debt crisis has been channelled away from the financial sector to the most vulnerable sectors of society. Debt repayment is made possible by sectors in social and economic hardship, not those individuals and sectors that feed back into the global financial sector via capital flight. The strategies established by development economists from multilateral institutions to recoup as much repayment as possible from financially restricted nation-states, are the antithesis of development, however constructed.

Alternative strategies must be implemented if Latin America's poor are not to pay for the lending institutions' mistakes of the 1970s. Whatever many and varied debt management scenarios are envisaged, it is an understanding of the wider operations of the global political economy that will provide directions towards an amelioration of the existing problems (Corbridge,

Table 3.10 Latin American external debt, 1984–94 (US$ billion)

	1984	1985	1986	1987	1988	1989	1990	19
Total external debt	398.3	412.5	434.8	474.9	456.1	452.4	476.5	49
Long-term debt	346.8	366.4	397.8	429.3	406.3	392.1	398.4	40
of which IMF	11.5	14.5	16.3	18.1	16.3	15.6	18.1	1
Short-term debt	51.5	46.1	37.0	45.5	49.9	60.3	78.1	8
Principal arrears on long-term debt	6.8	7.0	9.6	12.4	15.1	18.5	25.9	2
Interest arrears on long-term debt	3.3	2.9	3.7	8.6	9.0	17.1	26.0	4
Total debt service	53.5	48.6	49.1	47.1	55.3	51.5	46.4	4
Ratio of debt service to exports	38.3	36.6	42.5	36.7	37.5	31.0	25.3	2

Source: United Nations, World Economic and Social Survey, 1995

1992). The debt crisis is symptomatic of the nature of the historical development of the global political economy and the power relations established within that development process. In their analysis of the Latin American debt crisis, Stephany Griffith-Jones and Osvaldo Sunkel (1986, 132) note that:

> From the point of view of the governments of debtor developing countries (and particularly Latin American ones), debt crisis management has been fundamentally unsatisfactory, as it has not contributed towards sustaining, far less increasing, growth and development. The burden of debt servicing on Latin American economies in recent years and in future ones ... is still inconsistent with rapid economic growth.

1982 opened up a fracture in the path of external funding and resources towards Latin America. Payment of debt principal and capital service forced reductions in internal domestic spending as capital reserves diminished. On the trade front, imports were reduced and attention was focused on export-led development for the generation of foreign exchange. The impacts were not only on production and the standards of living of the majority of Latin Americans, but also on the Latin American capitalist accumulation capability (Eyzaguirre and Valdivia, 1989). The adjustment programmes and export strategies had to respond to the international crisis of higher interest rates, reduced trade values, the interruption of capital flows into the continent and the economic slowdown in the North. As with the 1929 'Crash', isolation from major markets and financial sources impacted heavily on the Latin American economies and signalled a shift in their political economies, a marked sea-change from their growth rates of the 1970s.[18] The same was also true in other continents, with a consequent impact on international trade (Vial Ruiz-Tagle, 1989).

Adjustment policies were formulated according to perceptions of the severity of the balance of payments problems faced. Broader economic policies, such as exchange policy, followed. The economic policies were responsive to international circumstances but the underlying domestic inflation and unemployment had not gone away. Economic attention was geared *hacia afuera* (outwards), with policies superimposed on existing problems but with different aims – Meller (1989, 79) calls it, 'the old conflict between trying to have an autonomous internal policy and being an active member of the international "economic" club'. The costs of adjustment have been transferred inequitably both socially and spatially as a result of policy formulations, as public sector jobs have been severely cut and the Latin American states have reduced resources directed towards urban and regional development administrations, agencies and projects.

The way out of the debt crisis has moved in the direction of 'adjustment with growth', the belief being that a growing economy will enable debt service. The policies have been poles apart from those adopted by the South-east Asian Tigers, where key sectors were targeted, investment channelled and special incentives (tax, credits) implemented. While comparison with the Jaguars and Kittens of Latin America offers little, since historical contexts and social and political structures are so different, one may make the point that austerity measures such as cutting the public sector have done little except shift the costs of adjustment to those social groups in 'peripheral' spaces (both urban peripheries and national territorial peripheries) (Jenkins, 1991). The expansion of the informal sector from the early 1980s speaks worlds for the failures of Latin American economic policies to distribute costs evenly (de Soto, 1989; Thomas, 1995).

The debt crisis shocked the Northern financial community. Overexposure to Latin American economies reduced bank profits since loan loss provisions set aside capital to cover their risk, while other banks were forced into closure. For Latin America, the effects of the mechanisms imposed on the economies to recoup their losses have been the opening up of the economies for new foreign investment in newly privatised activities and underpriced domestic operations. Beyond the macroeconomic context of the debt, the social effects of neoliberalism and the SAPs have been devastating.

The debt crisis established a geoeconomic power relationship between Latin American societies and multilateral lending institutions backed by Northern nation-states that curtailed any possibilities for sustainable development in its broadest political, social and cultural contexts (Griffith-Jones and Sunkel, 1986). This relationship has been exercised with little regard to the social impacts of the policies, revealing the fallacies embodied in the 'development policies' designed by these institutions. The net outflow of capital from Latin America to Northern states, firms and financial organisations reveals the underlying power relationship and its motivations and objectives.

The debt crisis not only shaped the geoeconomic relations of Latin American countries with other states, but also between states and private and multilateral financial institutions. In terms of the SAPs and other means of encouraging continued remittances, Latin American states have initiated new styles of economic activity, new styles of governance and regulation, and new styles of social management to accommodate the external financial pressures placed upon them. These new styles of organisation have had profound effects on territorial spaces, in terms of ownership and control of economic activities, and employment and migration patterns.

The power-space geometries that have emerged since 1982 have been distinctive and profound in terms of shaping Latin American socio-economies and spatial organisation. The political geographies of the debt crisis, inclusive of geoeconomics and geopolitics, are those of the reorganisation and restructuring of most Latin American spaces, urban and rural, during the 1980s and 1990s in order to accommodate the demands of global finance. The consequences of the debt crisis have led to the antithesis of social development in Latin America during the 1980s.

GEOPOLITICAL ECONOMY

From geopolitics to geoeconomics

In many Latin American states, the transition from the Cold War to the post-Cold War period has provided the international context for the transition from the predominance of post-Second World War authoritarianism to democratic political regimes. The extent to which these two processes are interrelated should not be underestimated. As an important region of Cold War-linked 'hot wars', and also of authoritarianism, terrorism and counter-insurgency to the left and right wings of the political spectrum, Latin American affairs were heavily influenced from beyond the region during the 1945–90 period. Investment, aid and military assistance flows bear witness to this influence. In terms of the activities of secretive organisations such as the CIA, KGB and their related branches, little information can be corroborated, but their stabilising and destabilising influences across the continent during the Cold War played a key role in determining Latin American political and economic strategies.

The post-Cold War period has been characterised by different influences as the international security agenda moves towards economic, social and environmental issues challenging the Cold War emphasis on political and military concerns (Buzan, 1991). The transition from political and military priorities to those of economic, social and environmental considerations questions the extent to which traditional geopolitical conceptions of state power, territoriality and sovereignty can enable an understanding of the 1990s global balance of power. Undoubtedly, there is an ongoing transitional shift of this nature within Latin America, with the issues of environmental degradation, AIDS, drug addiction and drug trafficking ascending the political agenda (Zaldívar, 1993).

The Cold War locked the greater part of the world into a geopolitical impasse. The USA and the USSR exerted political and economic influence around the world in order to conduct an ideological struggle manipulating surrogate nation-states, principally within the southern hemisphere, for direct opposition. The end of the Cold War has thrown up numerous debates about the nature of the global political economy, about which nation-states and which regions are predominant, and who or what is setting global agendas.

An important outcome of the academic discussions within the geography, political science and international relations literature on this theme is that the ideological political structuring of the Cold War era has passed, and, increasingly, that the role of economic forces is shaping the development of 'the global' (Thrift, 1992). To deal with this transition, geopolitical theorists have had to inject a more significant economic element into their work. This has led to political geographers stressing the importance of geoeconomics as much as geopolitics. However, these two terms do not sit well apart, and the two components have therefore been summed and referred to as geopolitical economy or 'geopolinomics' (Agnew and Corbridge, 1989, 1995). The term identifies the spatiality of the global political economy.

An important shift in the balance of the geopolitical (the power-space relations of political agents and processes) relative to the geoeconomic (the power-space relations of economic production and consumption) has been triggered by the globalisation of finance and production and the transfer of technologies that have challenged the centrality of the inter-state system in the global order. The new pressures on the state (rather than the fact of it being 'hollowed out') in the face of these global economic activities have given impetus to a corrective adjustment between geopolitics and geoeconomics (Thrift, 1992; Sideri, 1993).

The most significant advances in the study of these changes have been generated by international relations specialists in the reformulation of international political economy (IPE) beyond its orthodox traditions (Murphy and Tooze, 1991; Strange 1991, 1994a). These advances take IPE and move it beyond its preoccupation with nation-state activity to be inclusive of other sources of potential power, principally markets and technology. Geopolitical economy adds the spatial framework to the 'new' IPE.

The outcome of extending the points of reference ensures that flows of knowledge, security, production and finance emanating from these principal sources can be highlighted to establish changing power relations (Strange, 1991). Civil society, the roles and actions of citizens and groups of citizens, is a key factor that might be added to the framework to ensure that other flows that lead to power relations are also considered. As an important source of alternative political structures and power bases, movements emanating from civil society, rather than the conceptualisation of society as a market of consumption and little more, are required as an additional component. This is especially relevant in the case of Latin America in the 1990s.

Within the Latin American context, the advantages of a 'new' IPE over traditional geopolitical approaches can be found within the very nature of the state (Malloy 1989, 1993). One may suggest that the role of the state within geopolitics revolves around a notion of national interest (Dalby, 1991). Within Latin America this becomes a highly debatable and complex issue. In Central America for example, the state consistently fails to represent the interests of the majority within its sovereign area. State control slides from one unrepresentative interest group to another via the persistent failure of electoral politics and the influence of exogenous pressures, and the nations of the isthmus struggle to establish even a semblance of a democratic political system. Edelberto Torres Rivas (1981, 100) is not alone in his opinion when he states that the Central American state 'has turned inefficient and corrupt'.

The lack of a democratic tradition in the face of post-independence *caudillismo*, totalitarianism and bureaucratic-authoritarianism is a critical

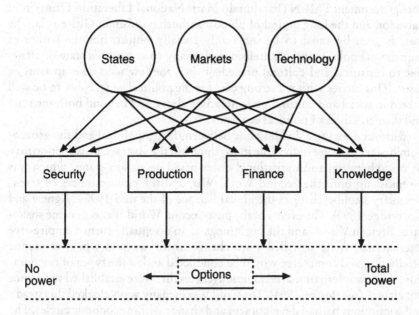

Figure 3.1 A framework for the 'new' international political economy (IPE)
Source: Strange (1991)

factor with regard to Latin American political economies, but more importantly issues of state power and action are most often centred around intra-state rather than inter-state affairs.[19] The role of the Central American state in internal affairs persistently revolves around repression rather than consensus. In international affairs these same states continue to be manipulated by US influence. Repression persists as the tool, while it is an arbitrary concept of 'international interest' rather than 'national interest' which is employed.

In view of these two features, the nature of internal conflict and the extent of external influence, which are inevitably interlinked, it is difficult to employ the same geopolitical tradition that emerged from the European 'Great Powers' with their state apparatus founded upon western liberal democracy. As such, new IPE offers a more flexible framework with which to work through Central American political geography. It is clearly evident that within the region political power has historically been wielded by those with economic power and that such economic power has been preserved by links with US companies, themselves linked to US government officials and departments.

The maintenance of a domestic entrepreneurial élite has served to concentrate wealth and polarise society. The outcome of such polarisation may be seen in the Sandinista revolution in Nicaragua, the activities of the anti-government FMLN (Farabundo Martí National Liberation Front) in El Salvador, and the long period of bloody authoritarianism in Guatemala. The state is seen by most to be inherently socially unjust, but the nature of patrimonial politics and clientelism leads many to accept this state of affairs due to historical and cultural precedent. It is the few who take up arms or resist. This carries with it a strong case for the political geographer to be well versed in social and cultural understanding above and beyond both the state and the necessities of political economy.

Without drawing a dividing line between the geopolitical and the geoeconomic, since they are indivisible in anything but the abstract, a distinction may be made between an international order based on nation-states, which was foremost up until the Second World War, and the circumstances of transboundary neoliberalism in the global balance of the mid-1990s (Agnew and Corbridge, 1995). The events of the post-Second World War economic shakeup at Bretton Woods and the beginnings of an essentially formal empire-free international arena signalled a sea-change in the ways in which economic production and commerce would be conducted and in the types of organisation, the new international economic agencies that were established to oversee this change (see Popke, 1994). The Cold War in many ways cloaked this steady and continuous path of development and change within economic circles. The cloak has now been removed and nation-states can be seen to have progressed very little in terms of their internal structures and mechanisms, and of their forms of governance and control.

The drawing aside of the Iron Curtain that had been hung by Churchill, symbolised by the fragmentation of the Berlin Wall into millions of post-Cold War souvenirs, has given rise to a transitional state of international affairs. The Cold War maintained a *status quo* in which states did not adapt and evolve to the degree experienced by firms and markets. Forms of governance were held in check, judged critically by the whole political spectrum, right to left, free market to command economy. Coinciding with this slow progress at the nation-state level, firms were responding rapidly to the B phase of the fourth Kondratieff wave and the Third Industrial Revolution (Taylor, 1993a; Gill, 1992). Hi-tech products, extensively used by the security forces and militaries of the superpowers and their respective partners and surrogates were only slowly taken on board within non-military circles, that is, within the civil bureaucracies.

As firms contended with the logistics of the internationalisation of production and of labour, a rise in intra-firm trade characterised and marked out a new transnational phenomenon (Dicken, 1992). The outcome was that world-wide more of the economically active population became employed by multinational firms, in a shift from state-controlled companies. In Latin America this is an ongoing process that has led to a rise in informal sector activities as the numbers of public sector employees decline rapidly and the privatised firms rationalise their workforces.

The 'geopolinomic' world

As a reflection on the 'geopolinomic' world, a disconcerting point is raised by Paul Kennedy (1993, 47) when he writes that multinational corporations have 'more global reach that global responsibility'. Maybe transnational forces hold the seeds of their own destruction and will mark a return to nation-state predominance and demands for control within a new regulatory-institutional framework, despite the inherent difficulties of over-all regulation beyond the national regulatory territory (Lipietz, 1987; O'Brien 1992). Until this happens, however, the rules of the global game are increasingly difficult to fix, and almost impossible to apply since firms are leading the way in 'the neoliberal spiral of regulatory degradation' (Peck and Tickell, 1994, 325), or 'systemic dislocation' (Leyshon, 1992, 251).

Geopolitical theorists have been forced to shake off their international shackles and acknowledge the unregulated and footloose world markets. The operations of the inter-nation-state system can no longer fully explain global processes. Late in the nineteenth century, Halford Mackinder envisaged the significance of the rapid increase in railway mileage in terms of the strategic condition of the Russian Heartland (Parker, 1982). A century further down the line, how does the information superhighway, or 'casino cyberspace', effect a better understanding of the interaction of the world's political and economic actors? This message may come as nothing new to those now engaging with various ideas of a 'new' or critical geopolitics (Ward, 1990; Agnew and Corbridge, 1989, 1995; Dalby, 1991; O Tuathail, 1992). Never-theless, where does it leave the *inter*national relations specialists – might they become *trans*national specialists? Clearly the message is one of global, not international, political economy, with a focus on transnationalism and uneven globalisation (Holm and Sørensen, 1995). As Stuart Corbridge (1994, 297) rightly points out, there is the possibility that 'Geopolitics will not only be funded by a world economy, it will be increasingly obliged to serve it.'

In summary, the inter-state system (the geopolitical) seeks to establish an embeddedness of clearly defined relations, and protocols for their effective maintenance. By contrast, actors within the geoeconomic seek to evade regulation, evolve constantly in response to competition, markets and technological adaptation, and redefine their relations in order to gain advantages. The late twentieth-century transnational arena is becoming a 'borderless' entity, as deterritorialisation and a seamless world are established by economic forces. While nation-states during the 1945–90 period dealt their geopolitical cards, of geostrategy and dominos, of georealms and rimlands, of shatterbelts and 'hot spots', firms unshackled themselves from national restrictions and obligations and established the idea of 'the global', as embodied in the multinational enterprise during the 1950s and 1960s, 'casino' capital during the 1970s (after the OPEC 'oil shocks') and 1980s entrepreneurialism (especially apparent in technology and forms of information as characterised by Rupert Murdoch (News International) and Bill Gates (Microsoft)).[20]

Latin American geopolitical economy in the mid-1990s is dictated on the formal level by the international financial community through the multilateral lending institutions. Geoeconomics determines democratic stability in the continent, except in the cases of Cuba, and, until recently, Nicaragua, where political ideologies still predominate. However, these are exceptions to the rule in the continent, as the neoliberal model, via SAPs, has been tailored to particular republics. The neoliberal experiment has resulted in profound economic changes in terms of private/public sector ownership and control over national territorial space. However, the drive for foreign exchange earnings via export-led development, investment and open markets has been met by a counter-culture of economic informality which is widespread.

The formal/informal split is serving to fracture Latin American societies. There are those people who live and work within the framework of the nation-state, and those who do not, existing 'beyond' the state. One outcome has been a continuation of the historical polarisation of Latin American societies. In the past this led to brutal authoritarianism and radical armed opposition but contemporarily it leads to a confrontation between weak democratic state administrations and popular grassroots movements. Another outcome has been the development of alternative economic strategies. The Latin American drugs industry is the most powerful example.

Box 3.1 The 'Drugs War': informal economics or narcoterrorism?

The Latin American drugs industry is based in the Andean regions where coca plants grow well. With demand high in international markets, particularly the USA (6 million cocaine users consuming 60 per cent of world cocaine) and increasingly Europe, illegal production and distribution have grown rapidly in many areas since the 1970s. Drug costs are kept high due to the risks of illegal production and this has led to large flows of capital into the production areas controlled by 'drugs barons' and 'drugs cartels', consortia of drugs production and distribution operations. With low income levels and relatively impoverished regional economies where production takes place, the large profits generated by the narco-business have resulted in drugs barons being highly influential within local areas, as employers, benefactors and social controllers.

Pablo Escobar and the Medellín cartel became the focus for Colombian and US authorities when the Colombian government targeted the drug problem.[21] Escobar was captured and imprisoned in a 'personalised' prison in 1991. The outcome was that the Cali cartel took up the slack in the drugs market. During this two-year period, coca production tripled in Bolivia as Colombia became the target for the US Drug Enforcement Agency (DEA) and the drugs barons sought safer production sites.

Apart from the personalist element of the drug barons, production aspects have had a great impact on many rural areas. The production of coca paste from coca leaf is profitable but at the same time illegal. This provides a

Plate 3.3 The role of large multinational companies has had a marked impact not only in production and labour activities but also in consumption patterns and cultural change. Intensive marketing has been integral to these changes; the term 'Coca-Colarisation' has been employed to explain this phenomenon and its effects. San Salvador (El Salvador).

Plate 3.4 In the late twentieth century, tourism has also played an important part in the generation of foreign exchange on the national scale, and as an important source of household income at the local scale. In many locations in Latin America, tourism (including the sale of artisanal products) has replaced other forms of economic activity as the principal source of household income. As with other forms of informal activity, children, like this girl in Antigua (Guatemala), are important income earners in this sector.

quandary for many local agriculturalists. Generally it is the influence, or threats, of the cartels on the coca producers that leads to many people pursuing its cultivation. Many areas of Colombia, Peru and Bolivia have become valuable productive economic spaces but the state has little control over this activity. Programmes of eradication and crop substitution, backed by finance and force from the DEA, have had little impact on the industry since the areas of production lie 'beyond the state', existing as cartel fiefdoms rather than state-controlled territories.

Another consideration is the role of drug profits in the funding of political movements and terrorist activities. Peru's Shining Path (Sendero Luminoso), for instance, has been active as an intermediary between coca producers and the distributors (*narcos*) in the Upper Huallaga valley, an area it controlled due to its relative inaccessibility and constructed as a quasi-state structure known popularly as the 'Huallaga Republic', in which it controlled coca prices, taxed *narcos* and enforced its own law and order. Another political armed movement, the Tupac Amaru Revolutionary Movement (MRTA), controls the central valley area. The Shining Path-controlled production is part of Peru's 65 per cent contribution to international cocaine output, a contribution that is the financial equivalent of half Peru's legal exports. Coca cultivation occupies approximately 200,000 hectares and approximately 300,000–400,000 people are directly or indirectly involved in operations. With high unemployment and underemployment, and coca workers receiving an income three times higher than domestic and menial workers, the 'coca-nomics' makes good sense to Andean rural workers. It is the link between 'coca-nomics', uneven development and poverty in rural areas, and the radical ideologies and economic activities of political movements such as Peru's Shining Path and MRTA that construct geographies of alternative intra-national spaces, controlled by non-state organisations and engaging in profitable, illegal economic activities. The question is the extent to which they feed one from another and how separable drugs policy and anti-terrorist policies really are.

Cocaine is not solely a Latin American problem. The geopolitical economy of coca cultivation is based on an international plane: supply and demand, cultivation and control, financing and laundering, all work within and beyond Latin America. The drug producers and production sites are 'footloose' and their geographical dispersion, from the initial concentration in Colombia into neighbouring states, has revealed the international complexities of control. As an informal, illegal economic activity, drug production has an economic geography that has had a major impact on Andean states, in terms of the governance of localities, of alternative income sources, and the alternative economics of the 'coca-dollar'.

Drug production is clearly tied up closely with low levels of economic development and restricted income streams in rural areas. Until rural economies enjoy better returns from traditional products, the coca leaf is unlikely to be eradicated and the 'barons' and terrorists will generate high incomes. However, to suggest that solutions to the drugs problem are to be found only in Latin America is to dissociate the demand side of the drugs equation, which takes us beyond the continent altogether: it is Uncle Sam who is the addict, the consumer of this highly valued Latin American product. New anti-drugs initiatives such as the 1990 Cartagena summit of the US,

Peruvian, Colombian and Bolivian presidents (which reconvened with the addition of Ecuador, Venezuela and Mexico in 1992), and the US pledge of $1.1 billion over five years to the Andean Initiative of economic aid to combat drugs, will continue to struggle in the context of an illicit trade war to supply the North American marketplace. There is evidence to suggest that the Incas controlled the use of coca consumption within their empire and that a black market in coca originated in the fifteenth century. Five hundred years on, coca cultivation remains unchanged except for the economic geography of its distribution.

Sources: Morales (1989), Gonzáles (1992), Hargreaves (1992), Poole and Rénique (1992), Doyle (1993)

An important point relating to SAPs and associated adjustment programmes carried out in Latin America is who is most affected within Latin American societies as the share of the cost of living is increasingly borne by families rather than government redistributive and social policies and programmes. The gender balance is very distinctive, for example (Nash and Safa, 1985). Over a longer-time scale, the female workforce has tripled during the 1950–80 period in Latin America and the Caribbean, with female participation rates increasing from 18 per cent to 26 per cent of women (Safa, 1995). These figures reflect the significance of women workers within Latin American economies, highlighting the concomitant changes within the home in terms of social and economic activities as women become more 'visible' in terms of the home/work, private/public, *casa/calle* dualities. Despite these changes, women continue to establish their social identity through the family and home while men do so through their economic role (Standing, 1989). An example of the changing gender balance is that approximately 20 per cent of Latin American households may now be defined as female-headed, often within a low income and single parent context (Chant, 1985, 1991). The impacts of these changes on the political geographies of the home and on changes in public spaces have been profound.

During the 1980s, the insertion of women into Latin American economies gained pace (Safa, 1995). The reasons for this increase are numerous but key features are the demand for women in the service sector and in export processing, and the gendered patterns of rural-urban migration. The flexible nature of women's work and their remuneration rates still suggests that their insertion into the economy has been a cheap fix in that their wages are lower and that the demands on women often increase as a result.

As bread-winners, family carers and community organisers (the activities in a woman's 'triple day'), women have been increasingly exploited in the neoliberalisation of Latin American economies, achieving greater social independence at the high expense of bearing many of the costs of the process. As women generally have become worse off, female-headed households have experienced the most critical problems. Tanski (1994, 1639)

Table 3.11 Rates of participation in economic activity (by sex, selected countries) (%)

Country	Total 1970	1985	1995	Women 1970	1985	1995	Men 1970	1985	1995
Bolivia	46.5	47.3	47.8	19.5	25.8	28.8	74.6	69.8	67.5
Colombia	44.5	46.9	49.6	20.2	29.7	33.2	69.6	64.6	66.4
Cuba	41.9	46.6	54.6	15.9	30.0	38.7	67.0	62.9	70.6
El Salvador	49.8	47.8	48.9	20.9	23.0	25.6	78.6	74.1	73.9
Haiti	72.7	58.4	58.1	65.5	46.7	46.6	80.4	70.7	70.3
Mexico	40.4	48.6	51.7	15.2	27.4	30.9	65.7	70.1	72.9
Peru	46.9	46.8	48.2	26.0	26.9	28.3	67.6	66.6	68.1
Venezuela	41.7	45.2	48.2	17.8	24.6	29.2	65.2	65.6	67.0

Source: CEPAL, Statistical Yearbook for Latin America and the Caribbean, 1994

notes that 'The "free" market is especially questionable when different sectors of the population have no freedom to choose how market forces will affect them.' Economic restructuring has weakened labour, and thus the social fabric, in the process of strengthening capital.

The *maquila* industries of northern Mexico provide good examples of women's insertion into economic production. In 1993, *maquiladora* plants in Mexico became the most important export manufacturers in the country. Their rapid development and economic success has been based on women's labour. As mainly non-unionised, low wage workers, women provide a compliant labour force, although the proportion of men now moving into these same non-unionised, low pay activities is increasing. Union membership in the *maquiladora* industries is 10 per cent relative to 25 per cent nationally, although there are some exceptions to this rule such as the significant umbrella organisations: the Confederation of Border Workers (CFO) and the informal Confederation of Women's Committees. The failure of the Mexican state to enforce its own gender and wage-related labour legislation results in effective state subsidisation of the *maquila* sector (Davidson, 1991; Kopinak, 1995).

In a case study of *maquiladores* in Ciudad Juárez (Mexico), María Fernández-Kelly (1983) makes the point that women's involvement in the labour force does not necessarily improve their alternatives as individuals or members of families since their employment in *maquiladores* may be a symptom of economic vulnerability and the lack of alternative employment opportunities. She adds that recognition must be made of the heterogeneity of the labour force, between the clothing and electronics sectors, for example, where labour demands differ in terms of age and skills, and also that labour must be viewed in terms of households and wider labour markets rather than solely in terms of capital accumulation.

By looking at diversity within the *maquiladora* industries and the connection between the workplace and home, the socio-economic context of

Plates 3.5 and 3.6 Women are often engaged in informal sales activities. The products they sell are often associated with food and clothing which can be seen as a commercialisation of women's traditional domestic activities. La Paz (Bolivia).

this manufacturing sector becomes more evident. Since it is this sector that has shaped the economic and social geographies of northern Mexico, and has had a profound impact on the Mexican economy as a whole, the political organisation of the region in terms of industrial location, trade and the gender landscape becomes highly significant.

After the Mexican debt default in 1982, the commercial credit flow to Latin America was sharply reduced and this resulted in a fall in raw material commodity prices. Many countries, such as Peru, which had undergone recent crises of production and inflationary pressures, leading to jumps in inflation figures, entered a critical phase with wide-reaching socio-economic repercussions. The international recession opened up the internal weaknesses and socio-economic fractures. In the Peruvian case, debt servicing absorbed $1 of every $2 dollars, leading to President García placing a 70 per cent limit on export revenue for debt interest repayment (Reid, 1984). In Ecuador, the imposition of a neoliberal economic team similar to Chile's 'Chicago Boys', known in Ecuador as 'The Dynamo', by President Ulloa secured a $960 million agreement with the IMF (Corkill and Cubitt, 1988). The choices available to Latin American economic teams were limited: follow austerity or risk isolation from the international financial community.

For the poorer countries in Latin America, it has been impossible to achieve any degree of political and economic autonomy in the face of the pressures of the international financial community. Bolivia, second only to Haiti in the western hemisphere poverty rankings, is a case in point. In the late 1980s, an externally devised recovery plan was implemented, called the New Economic Policy (NEP). Policies became linked to aid packages, and the Bolivian economy was opened up to international investment. Its most important sectors, such as oil and gas reserves and tin mining, were soon swept up by multinational investment. The tin sector was further hit by a price crash in the mid-1980s. The remainder of the public sector, 128 organisations, met with little external interest (Farthing and Villegas, 1991). Bolivia's example is extreme, but one that has been mirrored throughout the continent.

The two most important features of Latin American political economies in the mid-1990s are the shifts in economic orientation and the democratic transitions. Democratic transitions, from authoritarian regimes to civil democracies, have been taking place throughout Latin America since the mid-1980s, due to external and internal political pressures on regimes. International economic pressures have also played a part, such as the early 1980s international recession. While the timing has been variable within the continent, most nation-states currently find themselves working within the frameworks of both. The shift in economic orientation has been the shift away from the postwar focus on state interventionism in an economy buoyed up by the defence of protectionism, to economic liberalisation.

Liberalisation in its late twentieth-century guise, neoliberalism, has meant that domestic markets have been opened up for foreign investment in all sectors, from goods to services to finance. Deregulation by the state has given firms greater flexibility within Latin American economies, relatively untroubled by financially weak state organisations, as in the Mexican *maquila* industry (Campbell, 1991). The public sector has withered as its activities have been privatised, and falling levels of basic goods and of food security, and social polarisation have been the consequences (Barkin, 1991). Statistics of social change such as the United Nations Human Development rankings reveal a different picture of Latin American development from the one offered by macroeconomic data.

The state corporatism of the postwar period, within which the state

Table 3.12 UNDP Human Development Index[a] rankings

Country	HDI ranking 1992 (of 174 countries)	HDI 1992	1987	GDP per cap. 1992 (US$)	GDP per cap. rank/HDI rank[b]
High human development					
Costa Rica	28	0.883	0.916	5,480	32
Argentina	30	0.882	0.910	8,860	9
Uruguay	32	0.881	0.916	6,070	21
Chile	33	0.880	0.931	8,410	8
Venezuela	47	0.859	0.861	8,520	−7
Panama	49	0.856	0.883	5,600	10
Mexico	53	0.842	0.876	7,300	−6
Colombia	57	0.836	0.801	5,480	3
Brazil	63	0.804	0.784	5,240	1
Medium human development					
Ecuador	68	0.784	0.758	4,350	5
Cuba	72	0.769	0.877	3,412	16
Paraguay	87	0.723	0.784	3,390	3
Peru	93	0.709	0.753	3,300	1
Dominican Republic	96	0.705	0.699	3,280	−1
Nicaragua	109	0.611	0.743	2,790	−4
Guatemala	112	0.591	0.592	3,330	−20
Bolivia	113	0.588	0.548	2,410 '	−1
El Salvador	115	0.579	0.651	2,250	3
Honduras	116	0.578	0.563	2,000	4
Low human development					
Haiti	148	0.362	0.356	1,046	1

[a] Based on the UN Development Programme's calculations using measurements of life expectancy, literacy rates, schooling, and adjusted income. HDI values are between 0 and 1. The score 1 indicates higher levels of human development.
[b] A positive figure indicates that the HDI ranking is higher than the GDP per capita ranking; a negative figure indicates that the HDI ranking is lower than the GDP per capita ranking.
Source: UN Development Programme, *Human Development Report* (1995)

utilised a large share of GDP and was the principal national employer, has passed away. In most Latin American republics, the state now functions as an under-resourced regulator of private capital and its activities. Venezuela's recent failed coups and resulting political instability, and Mexico and Argentina's financial crises bear witness to the turbulence within the continent's political economies to which exposure to external financial pressures has led. The succession of protests against the restrictions of SAPs throughout the continent also bears testament to the difficulties being

Table 3.13 Households below the poverty line[a] (%), selected countries

	Total	Total urban	Metro-politan	Other urban	Rural		Total	Total urban	Metro-politan	Other urban	Rural
Venezuela						**Panama**					
1970	25	20	—	—	36	1970	—	—	—	—	—
1981	22	18	12	20	35	1979	36	31	27	42	45
1986	27	25	16	28	34	1986	34	30	27	41	48
1990	34	33	25	36	38	1989	38	34	32	42	48
1992	33	32	21	35	36	1991	36	34	32	40	43
Brazil[b]						**Costa Rica**					
1970	49	35	—	—	73	1970	24	15	—	—	30
1979	39	30	21	34	62	1981	22	16	15	17	28
1987	40	34	24	37	60	1988	25	21	19	22	28
1990	43	39	—	—	56	1990	24	22	20	25	25
						1992	25	25	22	29	25
Guatemala						**Honduras**					
1980	65	41	26	52	79	1970	65	40	—	—	75
1986	68	54	45	59	75	1986	71	53	—	—	81
1990	—	—	—	—	72	1990	75	65	—	—	84
						1992	73	66	—	—	79
Mexico						**Chile**					
1970	34	20	—	—	49	1970	17	12	—	—	25
1977	32	c	c	c	c	1987	38	37	33	40	45
1984	34	28	d	d	45	1990	35	34	30	38	36
1989	39	34	—	—	49	1992	28	27	22	30	29
1992	36	30	—	—	46						
Latin America						**Peru**					
1970	40	26	—	—	62	1970	50	28	—	—	39
1980	35	25	—	—	54	1979	46	35	29	41	65
1986	37	30	—	—	53	1986	52	45	37	53	64
1990	39	34	—	—	53						

[a] Includes households below the indigence line or living in extreme poverty.
[b] Metropolitan area: average of the figures for Rio de Janeiro and São Paulo.
[c] Information available only at the national level.
[d] Estimates could not be made for the Federal District because of the small sample size.
Source: ECLAC, Social Panorama, 1994

experienced as a consequence of international economic interventionism (Green, 1995).

Geopolitically speaking, there are also significant territorial issues that become problematic under privatised economies. With national integration dependent on transport and communication links, privatisation threatens peripheral spaces and regional development policies. While state actors may function as regional developers, private actors have no specific regional development function beyond their own infrastructural and labour requirements. In Argentina's Patagonia region, the future development of previously state-controlled activities such as the petroleum, gas, iron, coal and hydroelectricity industries, now increasingly in private hands, will determine local and regional change, migration patterns and national territorial cohesion (Saint Lary and Colantuono, 1993).

As the private sector has taken over economic sectors *en masse*, markets have been 'freed up' from state manipulation, such as protection and wage and price fixing. However, there is the question of the extent to which the state can influence its own domestic economy since so many of its basic factors are beyond its reach within the private sector. To what extent are Latin American states becoming overseers and inefficient regulators of the productive and servicing activities operating within the national territorial space, rather than active players within these activities? Noam Chomsky (1993) argues that multinationals and governments are closely wedded and that governments insulate firms from public interference in their decision-making, ensuring that power and wealth remains within these firms and state élites. By adding the supranational institutions (such as the IMF, World Bank and G7) to the equation of national governments and multinational firms, Chomsky (1993, 64) suggests that the Cold War balance of power has been replaced by 'The new imperial age [which] marks a further move towards the authoritarian extreme of formal democratic practice.'

Depending on the political and economic leaning of the respondent, the answer to the enigma of the role of states in economies will vary. Between the extremes of Cuba and Chile, Latin America provides the full spectrum on which to base an answer. What is clear is that an understanding of the answer requires a wider remit that an analysis of the state alone. Geopolitical economy provides that broader framework within which the promotion of an equity-led growth model could mark a shift from the preoccupation with globalisation to one of social democracy (Roxborough, 1992). Within such an ideological shift, the redistribution of income would not move away from the poorest groups in Latin American society with low incomes and little employment security (if at all), but rather towards them. Duncan Green (1995, 121) promotes similar ideas:

The debate on alternatives is at present dispersed and commands little attention outside the region. Its clothes are regularly stolen by the

World Bank, with its renewed stress on education, poverty alleviation and redistribution, although always within a programme of market reforms and an increased assault on labour safeguards. Nevertheless, it could contain the seeds of a future model which seeks to correct the social injustice and poor macroeconomic performance of neoliberalism, delivering the elusive combination of growth with equity.

Clearly, geopolitical economy provides a useful framework within which to analyse the impacts of neoliberalism within Latin America, and the opportunities for alternative development patterns. Geopolitics, with its nation-state focus, has to confront the role of geoeconomics in the organisation of spaces. Once connections have been made, geopolitical economy constructs an overview of the power-space relations between political, economic and social actors and processes, and how they impact on one another to shape spaces.

Geopolitical economy provides a more complex perspective on spatial analysis. Partial perspectives on Latin America's most important obstacles to development, such as high levels of absolute poverty, have not led to effective paths of understanding and new initiatives. The failures of the IMF austerity policies and similar government measures to provide basic needs for Latin America's poor and the riots that these engender speak volumes in this respect. The geographical dispersion of these policies and riots since 1990 also says a great deal about the blanket implementation of this style of programme (Green, 1995): Argentina 1993, 1995; Bolivia 1990, 1992, 1993; Brazil 1990, 1992; Cuba, 1994; Dominican Republic, 1990; Ecuador, 1991; Mexico, 1994, 1995; Nicaragua, 1990; Panama, 1992; Peru, 1991; Venezuela, 1991. The impact of austerity has been continent-wide and devastating in its social impacts. However, within the context of geopolitical and geoeconomic relations developed in post-independence Latin America, it is little more than another face of the globalising forces of international capital. Latin America's geopolitical economy is a reflection of the demands of this capital rather than the demands of its own citizens.

FURTHER READING

Debates relating to geopolitical economy and the post-Cold War can be found in Agnew and Corbridge's *Mastering Space: Hegemony, Territory and International Political Economy* (1995), Demko and Wood's (eds) *Reordering the World: Geopolitical Perspectives on the 21st Century* (1994), and Lowenthal and Treverton's (eds) *Latin America in a New World* (1994). Historical political economy is provided in Skidmore and Smith's *Modern Latin America* (1989) and John Sheahan's *Patterns of Development in Latin America* (1987) while James Dietz's (ed.) *Latin America's Economic Development: Confronting Crisis* (1995) and Duncan Green's *Silent Revolution: The Rise of Market Economics in Latin America* (1995) provide contemporary analysis.

4

THE ORGANISATION OF CONTEMPORARY LATIN AMERICAN SPACE

> In both North and South America, 'American democracy' with its principle of self-government is partly a result of there being so much space. Space moderates everything and conserves everything – until it is conquered.
>
> (Braudel 1994, 432)

THEORISING ON THE LATIN AMERICAN STATE

The origins of the Latin American nation-state

The Latin American state, as a recognised component of the inter-state system, was a product of the struggles for independence in the early nineteenth century. Based on the European model that had influenced the leaders of the independence movements, these nascent states exhibited all the institutions and characteristics of the European nation-states at the turn of the nineteenth century, but it was the post-revolutionary state in France that provided the most influential example for post-revolutionary Latin America. Under the continuing firm control of the military leaders of the anti-Iberian struggle, the notions of freedom, brotherhood and equality became strong components within the constitutional and nationalist make-ups of the new states, although the extent to which these were realistically pursued is debatable. As in France, the independence day celebrations of Latin American countries remain the key dates in the calendar. Furthermore, the significance of the independence struggles are recorded in the names of the arterial avenues of the largest cities and by the principal public statues commemorating independence heroes. As a basis of nationalist sentiment and aspiration, independence and historical events *por la patria* remain as the building blocks of national identity.

From these post-independence beginnings, the Latin American states suffered from social polarisation as a legacy of landownership and economic practices established during the colonial period. The independence struggles

120

were not class struggles but anti-colonial struggles and the new states, divorced from their external colonising powers, began life with internal contradictions and destabilising internal dynamics. In this way, independence did not mark a break with the past of colonial subordination and expropriation, but signalled the emergence of new élites.

The domestic élites of European origin organised the democratic state as an extension of their political and socio-economic predominance based on landed wealth and commercial activities. In their pursuit of power they were strongly linked with foreign commercial interest groups. These groups established a post-colonial bourgeoisie of considerable influence due to their economic strength. A tight grip on national politics and the fledgling state apparatus was also achieved. With all the outward signs of the Northern state apparatus, the Latin American nation-states and the power relations that were established, in terms of ownership of land and productive activities, reflected the economic structures of the republics that the colonial powers had instituted prior to independence. The Latin American nation-state was therefore shaped by the colonial and post-colonial expansion of international capitalist relations and the power-space relations of capital embedded within them.

The outcome of the relative autonomy of economic élites in the nineteenth-century states led to regional influence around *caudillos* who exerted political power over localised areas. *Caudillismo* and the support bases that were established around these individuals set in motion the cult of the individual that is reflected in Latin American state theories of populism, and also the associated influences of paternalism, clientelism and authoritarianism. As opposed to European and US liberal democratic state theories, Latin America during the post-independence period has revealed the important role that individuals have played in the evolution of the political process. Much of the background to the prevalence of populist politics and authoritarianism based on key personalities, such as Perón and Velasco Ibarra (populists) and Castro and Stroessner (authoritarians), can be explained by the early development of Latin American republican politics and the relationship of this to the regional fragmentation of economies and their control. These developments can, in turn, largely be explained by the emergence of productive regions during the colonial period.

In many cases regions were linked with different centres during the colonial period, such as the north of Argentina with the Vice-Royalty of Peru rather than with the River Plate (Piel, 1989). Not until the turn of the century, with the growing influence of foreign capital and the monopolisation of land ownership within Central America, and industrialisation within Brazil and the Southern Cone, did nation-states exhibit a degree of territorial cohesion. This cohesion was facilitated by rail and road infrastructural developments during the late nineteenth and early twentieth centuries. Geographies of production and distribution and the ensuing (re)organisation of Latin

American space were significantly altered by the arrival of rail, and later road, transport, and steam-powered merchant fleets (Lewis, 1983; Downes, 1992).[1]

In terms of the formation of Latin American nation-states, recognition needs to be given to the numerous sources of influence within the continent that have given rise to heterogeneous nation-state formations. The varying stages and integration of pre-Columbian formations, of those of the conquest and colonial period, and those of post-independence republicanism make comparisons difficult in terms of differentiation in the state structures and functions through which state power is channelled (Valle, 1983–84). These processes have also led to differences in spatial organisation, from divisions set in place by pre-Columbian societies, and latterly from the imperial administrative units such as audiences, captaincies and vice-royalties.

A differentiation that may be made between different state developments, although there are clearly amalgams, is between Afroamerican, Indoamerican and European Latin American influences (Torres Rivas, 1981). A recognition that Latin American nations are not original is a starting point. They are, rather, amalgams of various overlapping state structures and power frameworks. As such, Latin American nations are not underdeveloped, which could be said if they are compared with traditional European liberal democratic models, rather they are distinctive.

The difficulty in Latin American nation-state formation has been in the integration of societies and ethnically diverse groups within a larger whole. The struggles for independence resulted in independent territories but not in the integration of peoples within these territories to achieve a common identity. As such, independence did not bring with it a contemporary nation-state. The nation-state has evolved as debates surrounding the power structures of independence have been fought over and negotiated. The outcome has been political cleavages leading to authoritarianism and radicalism.

Samuel Huntington's (1968) work on political change highlights the revolutionary characteristics that appear in Latin America. Quite simply, revolutionary change takes place when popular participation mobilises beyond and around formal political institutions, challenging the ideology and legitimacy of these institutions. The forces of the state are then engaged in opposition to this political threat. This situation has been reached many times in Latin America. In many of the cases, revolutionary activity came as a response to political exclusion under authoritarian systems, such as in Nicaragua and Cuba.

The Latin American left

The construction of a Cuban state along communist lines from the end of the revolutionary conflict in 1959 marked a turning point for the Latin American political left. Cuba became the focus for socialist and communist parties and movements throughout the continent, the model for an alternative form of state and organisation (Slater, 1986). While this alternative model and its promotion in other Latin American states resulted in Cuba's expulsion from the Organisation of American States in 1962, Mexico maintained relations with Cuba. Mexico's own revolutionary heritage gave this relationship some common ground but the states were very different ideologically, although similar in organisational terms due to the close connection of the state with the party. The Mexican Institutional Revolution Party (PRI) has drifted towards the right since the revolution of the 1930s while the left-wing Cuban Communist Party has kept hard to the left until some relaxation in the 1990s. However, both have exerted authority over the political system, over the state apparatus and throughout civil society in their particular national spaces.

The collapse of Marxist-Leninism in Europe has had a marked impact on Latin American radicalism, resulting in ideological shifts towards market socialism. However, extreme left-wing thinking still remains in many Latin American republics, acting within political systems and beyond them through armed struggle (see Castañeda, 1993b). Peru's Maoist Sendero Luminoso (Shining Path) provides the most influential example of the latter. For Cuba, the example of a Latin American political alternative, the 1990s have been characterised by the need to take a different development path. Oil consumption has been reduced as subsidised supplies from its former Communist partner states have dried up, tourism has been redeveloped to generate foreign exchange, and food self-sufficiency has been promoted. In terms of other important reforms, foreign capital has been attracted once again, labour reforms have been implemented, and processes of decentralisation of decision-making have been put in place. On the political front, constitutional reforms and new electoral laws will make sweeping changes of the political system that has remained so rigid since the 1960s (Edelstein, 1995). Cuba is undergoing a transformation that has signalled an end to the brand of communism of the Cold War period.

For Cuba, as for the rest of Latin America, the post-Cold War period has led to transitions that require a reconceptualisation of socialism and democracy in the continent (Petras and Morley, 1992). The left wing in Latin American politics has been slow, on the whole, to adapt to these changes. For instance, the third meeting of the Latin American left-wing forum (founded in Brazil in 1990) in Managua in 1992 failed to propose strong alternatives to the neoliberal surge in Latin America (Green, 1996). Without viable political alternatives on the institutional agenda, it is the social organisations and popular movements of Latin American civil society which will provide the principal opposition to the predominance of traditional political systems.

Box 4.1 For the Shining Path of José Carlos Mariátegui

In 1980, Abimael Guzmán, the leader of Sendero Luminoso (Shining Path), ended a speech entitled 'We Are the Initiators' with the words 'The World Revolution has entered the strategic offensive ... The future lies in guns and cannons ... The armed revolution has begun ... Let us initiate the armed struggle!' Fifteen years later, Guzmán is in jail, 25,000–30,000 lives have been lost in the armed struggle, and Peru has struggled socially and economically in the face of civil war.

The armed struggle, or 'People's War', began in May 1980 when supporters burned ballot boxes in opposition to the institutional political system. The movement started within the University of San Cristóbal de Huamanga in the southern region of Ayacucho with three key assumptions: that the peasantry would be the social basis of revolution; that violence played a fundamental role in the revolutionary process; and that Gang-of-Four brand Maoism was an invariable 'truth'. The goal was the transformation of Peruvian society from its 'semi-feudal and semi-colonial' condition to the 'Republic of the New Democracy', a peasant-worker republic headed by President Gonzalo (Guzmán's *nom de guerre*) which would turn to traditional agricultural practices and exchange and reject capitalism. Barter would replace the capital money economy, and modern techniques would be replaced by pre-conquest agricultural techniques.

Although the movement took Mariátegui's name, *indigenismo* ('Indianism') played no part in the movement's ideology, despite the fact that the Andean peasantry were more attached to a culture of messianic or millenarian movements than to European-based ideologies. One of the few concessions to indigenous culture was the use of the Quechua word *misti* (white outsider) to refer to foreign imperialists. Andean Maoism was essentially radical authoritarianism and as such was rooted in the Peruvian political tradition of authoritarianism.

By 1984 Sendero had established themselves in rural areas throughout the country, although the south was their stronghold. The theory behind Sendero activities was based on the *foco* concept of control of rural areas, strangling urban centres logistically.[2] This had been Mao's model and had also played a part in the Cuban revolution. Sendero was especially successful in the southern provinces (Ayacucho, Huancavelica, Apurimac), the poorest in Peru, where they filled a power vacuum left by national political parties beyond the urban areas. The state's reaction was the declaration of states of emergency in various provinces of the country, in President Belaúnde's 'holy war against terrorism'. The war against terrorism had started. While the state began to militarise the areas of greatest Sendero support, Sendero continued their *foco* activities, organised into three stages: (1) armed propaganda and sabotage operations, aimed at the consolidation of areas of influence in the countryside, the destabilisation and polarising of the political system, and the forging of an experienced guerilla army; (2) the establishment of liberated zones; (3) a final stage of total war and the surrounding and over-running of the cities from the countryside. The armed forces' response to the *foco* strategy was the forced removal of peasants from the highlands to 'strategic' villages on the lower slopes. This strategy was also used in Guatemala as a mechanism of control via the physical concentration of possible supporting communities. The outcome was that communities became polarised with respect to the armed struggle, leading to inter-community conflict.

Since the start of the conflict, Peru's economy has suffered immeasurably. Collapse in rural areas, the destabilising environment of the coca trade, and civil war between Sendero (and other revolutionary movements) and the community defence committees, the *rondas campesinas*, have all promoted instability. Sendero shows of force against urban areas such as attacks on infrastructure and 'blacking out' Lima by cutting electricity supplies, reveal the continuation of the *foco* strategy. However, Sendero has failed to mobilise any but a small percentage of the peasantry. Many more have been disillusioned by the violence of the movement and the scale of bloodshed. A return to peace is the most desired outcome, but the military are no closer to ending Sendero's campaign, and similarly Sendero look unlikely to mount a serious attack on the state beyond bloody terrorist actions.

Important in terms of the social organisation of Sendero is its gendered make-up. More women are active in the movement than men, and the movement has its heroines such as Edith Lagos, 19 years old when killed by the military, having been the first commander of the Ayacucho region (30,000 attended the funeral), also Brenda Pérez Zamora, second in command of Sendero since 1988.[3] With many men having left the highland areas in search of work in mining areas, in manufacturing and commercial farming since the 1960s, women's roles have changed. As the defenders of communal land and the cultural integrity of communal life, they have been easily mobilised by the rhetoric of Sendero. Another consideration is the historical association of Sendero with feminism. As early as the 1970s, Sendero had its own Popular Feminist Movement which made ideological connections between left-wing ideologies and the feminist debates.

Sources: Guzmán (1980), Andreas (1990–91), Strong (1992), Degregori (1992), McClintock (1992), Poole and Rénique (1992)

Bureaucratic authoritarianism

Contrasting with the Latin American experiences of revolutionary Marxist-Leninism-Maoism, right-wing authoritarianism has predominated within Latin American political history. Authoritarianism has had a profound influence on the organisation of state apparatus and its function relative to civil society, and on economic strategies and the associated organisation of national territory (see Stepan, 1973; Phillip, 1985).

Guillermo O'Donnell's (1973) model of bureaucratic authoritarianism provides the framework within which the majority of authoritarian states have functioned, whether dictatorial or oligarchic. The bureaucratic-authoritarian state seeks to control social and economic activities within a political system with limited forms of representation. This style of corporatism ensured the military supervision of all activities within the national space, even when these were private sector activities (Wynia, 1990). What must not be underestimated is the support that these styles of control generate. Many low income groups in Latin America are strong supporters of conservative populism. This helps to explain the resurgence of Bonapartist figures within authoritarian and democratic regimes. Peru's President Fujimori provides a contemporary example of such personalism. The outcome of this tradition is that presidential behaviour often leads to a weakening of party systems (Roxborough, 1992).

The long duration of late twentieth-century authoritarian rule in many Latin American republics, particularly Brazil and in the Southern Cone (Argentina, Chile, Uruguay, Paraguay) was founded on the national security doctrine of the threat of internal instability and on the maintenance of states of emergency, so-called states of exception (which became the norm). In this way, constitutions became suspended, granting authoritarian leaders overwhelming powers within the nation-state. In the case of Chile, General Pinochet installed a new constitution to support his rule in 1980, which was ratified by a plebiscite passing the measure with 67 per cent of the vote. The intimidating political climate within which the referendum was carried out explains the result to a large degree.

The transitions to democracy set in motion from the early 1980s, as the bureaucratic-authoritarian states struggled against the debt crisis and international recession, have led to alternatives to the extreme left- and right-wing systems of state organisation prevailing in the postwar period. Very much as

Plate 4.1 La Moneda, Chile's presidential palace in Santiago, was attacked by military aircraft during the September 1973 *golpe de estado*. The bombing of the centrepiece of Chilean constitutional politics revealed General Pinochet's disregard for electoral politics and democracy.

Plate 4.2 Civil-military relations in Latin America continue to be problematic, despite democratic transitions. A high-profile military presence still prevails in many areas of Latin America. Military base, northern Guatemala.

a condition of transfers of power from military to civil administrations, the economic strategies of the postwar period, such as the attraction of foreign investment and integration within the capitalist world economy, have prevailed. The neoliberal models have also been conditioning factors in the ability of the new democracies to establish alternative economic plans and consolidate electoral politics (see Petras and Morley, 1992). An important shift in this respect has been the rise of technical politicians, *técnicos*, rather than the traditional political party 'bosses', *políticos*. Mexico under Salinas and Zedillo, Argentina under Menem, and Chile under Aylwin and Frei, have revealed the econocentrism of the new democratic administrations. The control of inflation, the tasks of liberalisation and state privatisation are examples of the prioritisation of economic stability and an acceleration into the global economy, above and beyond the traditional politics of populism, clientelism and party alignments.

ELECTIONS, POPULISM AND PERSONALISM

The failure of electoral politics

While the administration of liberal democratic states in Western Europe and North America is decided by electoral politics (see Dunleavy and O'Leary, 1987), the political history of Latin American states reveals that democratic

THE ORGANISATION OF LATIN AMERICAN SPACE

transitions of state power from one executive to another have been less common in post-independence history than alternative forms of transition, be it a *golpe de estado*, *autogolpe* (self-coup) or regime prolongation via constitutional amendment or the application of emergency powers. Democratic executive transfers and electoral processes struggle against this historical legacy. In Ecuador for example, José María Velasco Ibarra was president five times (1934–35, 1944–47, 1952–56, 1960–61, 1969–72). The strong personalism and populism of *velasquismo* meant that the twelve years of democratic transitions of power between 1948 and 1960 have become known as the 'democratic parenthesis' (Corkill and Cubitt, 1988). Such democratic parentheses are commonplace in Latin America's political history, sandwiched between political instability and undemocratic political regimes.

For these reasons, electoral politics in the Latin American political geography context must be considered in historical and cultural context. The cultural context in this regard is the cult of personalism that follows from *caudillismo* and the significance of regional identities and local power bases within political systems. An extension of this is the impact of populist politics within the continent and its ability to mobilise large sections of the electorate around local issues, allocating votes to a candidate in return for social and infrastructural development initiatives. Latin American elections are still dominated by the élite and sectoral interest groups' influence over state bureaucracies, parties and the military.

In Mexico, the PRI's fear of losing power in the 1988 election campaign was overcome by a two-week delay in the announcement of the result, blamed on a computer problem, after which time the PRI was returned to power once again. In Paraguay, after the 1989 *coup d'état* when General Stroessner was removed from power, Stroessner's Colorado party support base continued to dominate Paraguayan politics, controlling the Central Electoral Commission, and thus the election machinery. The Colorado party was, not surprisingly, returned in the May 1993 elections (Knippers Black, 1993). In Argentina, President Menem ensured himself a second term of presidential office by making a pact with the Radical opposition party and amending the constitution. In Brazil, President Collor de Mello was indicted on corruption charges in 1992. In Chile, an unlikely broad-based coalition of parties from Christian Democrats to Socialists has succeeded in providing an opposition to a return to authoritarianism since 1989 (rather than establishing an effective multi-party competitive system). In Peru in 1992, President Fujimori succeeded in delivering an *autogolpe*, dissolving Congress then reconstituting it, in order to enable him to bypass the elected executive bodies to push through austere economic policies.

In very few cases in Latin America does one find successful democratic and legitimate transfers of executives based on the electoral process. Electoral histories have been tainted due to their abuse by authoritarian regimes to further consolidate their power (for example, the Chilean constitutional referendum of 1980, and Paraguayan elections under Stroessner) or provide a democratic facade to the outside world (as in Brazil, 1964–85). Another important influence in the electoral history of Latin America has been the role of external influences and support. In 1990, the US-backed campaign of Violeta Chamorro and her National Opposition Union (UNO) succeeded in ousting the Sandinistas from power in Nicaragua. In the same year in neighbouring Costa Rica, the National Liberation Party of President Oscar Arias was defeated by the pro-US candidate supported by the US National Endowment for Democracy fund.

In response to failures in the electoral process, movements based within civil society have confronted the inadequacies of institutional politics based on parties, elections and legislatures that continue to provide few benefits (Pereira, 1993). Jan Knippers Black (1993, 545) describes Latin American electoral systems as 'input' democracies in that popular participation in elections is not reflected in popular benefits from public policy. The suggestion is that Latin American societies are clearly neither European nor North American, despite the colonialising influences of both. Methods of transactions, political activity, and the use of influence and power have developed in distinct ways. For this reason, the practices of clientelism and patrimonial politics are accepted as integral to Latin American social and political life. This high degree of informality and interpersonal linkages may be translated as corruption and bribery in certain contexts, yet within the

Plate 4.3 Political graffiti are a common form of political expression throughout Latin America. This scene depicts a politician attempting to convince a worker to re-elect him: Politician: 'Vote for me and you'll be fine.' Worker: 'Is he mad? He didn't vote for legislation in my favour for five years.' Montevideo (Uruguay).

Plate 4.4 This graffito is a parody of the state of fear in Uruguay during the authoritarian regime (1973–85). Written by a youth wing of the Worker's Party, it reads: 'Manual for Suicide: (1) go out at night (2) go dancing (3) kiss your girlfriend/boyfriend (4) look at the stars (5) picnic in the *plaza*.' Montevideo (Uruguay).

Latin American social consciousness there is a clear concept of the relationship between social and political change and personal relations, in preference to institutionalism. The impeachment of Brazil's President Collor de Mello for corruption, on the one hand, and the mechanisms of patrimonial politics that are so important within Brazil's political process, on the other, demonstrate the fine lines that may be drawn.

Personalism and populism

The reduction of political, economic and social affairs to the personal is a prominent aspect of much of Latin American life. This may be seen as a mechanism to avoid the onerous bureaucratic systems operating within the continent, or as a connection with an historical tribute system of the pre-Columbian period such as the *mita*. Whatever the background, the role of the personal is key to an understanding of Latin American political geography. The influence of personalism has been prominent from the *kuraka*, *caciques* and *caudillos* to more contemporary figures: Perón, Castro, Pinochet, Rigoberta Menchú and Chico Mendes.[4] Each of these individuals, with their personalised political geographies, has transformed the political landscape to a considerable degree. As such they have undermined traditional concepts of

130

overriding state power and control. By exploiting emergency powers, suspending constitutions and stimulating popular opinion, they have each changed political spaces: spaces of political activity, of representation, and of freedom from constraint.

While contemporary individuals such as Rigoberta Menchú (see page 166) and Domitila Barrios de Chungará are important, in terms of indigenous rights and highlighting the political roles of women, there are also important historical references within Latin American political life to previous influential personalities. Many twentieth-century political movements make historical connections: the Sandinistas in Nicaragua (after Sandino, who opposed the USA-Somoza political system), the Zapatistas in Chiapas (after Zapata, the Mexican revolutionary), the Tupamaros in Uruguay (after the Inca leader Tupac Amaru), Lautaristas in Chile (after Lautaro, the Mapuche leader), the FMLN in El Salvador (after the revolutionary Farabundo Martí), and Peronism in Argentina.

Personalism, linked with *caudillismo* and pre-Columbian and colonial forms of patrimony, goes back to hero idolatry and subordination. Linked with traditional structures of semi-feudal, semi-capitalist structures of landlords and *inquilinos*, sharecroppers and tenant farmers (all forms of tied rural agricultural labour), Latin American social structures have been heavily imbued with power hierarchies. Popular movements at the grassroots level have marked an important alternative to the predominance of personalism, replacing it with greater self-dependency in political dealings within communities, especially where political surrogates are concerned.

Another issue relating to the power of individuals within political systems has been the action of a self-coup or *autogolpe*. Through this distinctive Latin American mechanism, individuals have maintained themselves in power in opposition to constitutional legislation and elected representatives, by exercising exceptional presidential powers. The *autogolpe* of President Fujimori in April 1992 enabled him to make wholesale changes within the government by dissolving the existing Congress and establishing a new administration. This administration was favourable to the second stage of the process known as the 'Fujishock', a series of harsh economic reforms that have raised Peru's popularity within the financial markets but had crippling social effects within Peru's socio-economy.

In the same way that the Latin American state structures are distinctive, the territorial-administrative formations of contemporary Latin American states are as diverse as they are complex. Brazil, for example, occupies half the Latin American continent and retains a federal structure based on the recently constructed and geographically centralised capital of Brasilia; Uruguay, on the other hand, is the size of Brazil's smallest states and exhibits a high degree of centrality within the capital city of Montevideo, functioning similarly to an Athenian city-state.

Despite the wide variation, the relations of power within national spaces have predominantly been of a core–periphery nature, with power emanating from decision-making conducted within the centre rather than from the periphery (Gottmann, 1973). This basic framework explains much of Latin America's territorial development. The roots of this situation may be found in the centralised systems of the pre-Columbian states, and in the vice-royalties, audiences and captaincies of the colonial period. Contemporary examples of the highly centralised nation-state based on the primate capital city may be found throughout the region, as in Mexico City, Santiago de Chile, Buenos Aires, and Tegucigalpa (Honduras).

Centralisation and decentralisation

The history of the Latin American state is one of authoritarian mechanisms of control. This has led to the concentration and centralisation of power within national capitals and key provincial production, commercial and administrative centres. At the same time this arrangement has extended authority outwards over the national space and led to a reduction in the regionalist tendencies and local power bases inherent within most republics. Any concessions to decentralisation during authoritarian regime control were generally mechanisms for the establishment of regional structures of military governance, as in Chile. In the post-authoritarian transitions of the 1980s and 1990s, forms of decentralisation have been promoted so as to develop local and regional initiatives and strategies in order to combat location-specific development issues.

Urban primacy resulting from the rush for industrialisation from the 1960s and 1970s and mass migration from rural to urban areas provided a *raison d'être* for cities to provide national focal points, administrative centres and political cores, especially the largest cities. In the case of Argentina, the centrality of Buenos Aires in the activities of the nation-state is noted as being the key to many of the problems associated with the organisation of Argentinian space (Gómez Insausti, 1988; Roccatagliata, 1988); the same criticism can be levelled at many other states. Despite this retention of power within large urban centres, there have been efforts to decentralise power prior to the 1990s. These have taken place in opposition to the continental political norm of centralisation and have resulted from the revolutions in Cuba and Nicaragua. The degree to which this decentralisation was successful or given full rein is debatable. However, institutions and structures were put in place to accommodate this centrifugal political process, filtering down to decision-making at the most localised levels of communities and workers' committees. While it is an over-simplification to note that author-itarian control requires centralisation and socialist or communist control can operate in a politically diffused environment, the issue that is raised in 1990s Latin America is the degree of (de)centralisation exhibited by the contemporary democratic regimes.

There is little doubt that there is considerable opposition to the centrality of the largest cities, in terms of government policy and resource provision, among those within the provinces, regardless of whether they live in urban or rural areas. National life is seemingly conducted within the largest cities and as such it demonstrates centripetal forces in terms of economic activity (especially foreign companies), migration, domestic bureaucracy and the media. While the issue of counter-urbanisation would not appear to be a political priority, the question of extending competent governance, linked with resources, beyond these centres certainly is. The insurrection in Chiapas (Mexico) provides a potent example of the failure to address the decentralisation issue, as do the activities of Sendero Luminoso in Peru. With decentralisation, social actors are able to participate in decisions relating to their local development, as well as impacting on regional and national decisions via the electoral process. In effect, the popular movements that sprang up around Latin America from the 1970s were all actively engaging in just such a process, yet they existed 'beyond' the state rather than as a constituent part of its hierarchical structure.

In Venezuela in 1984, the Presidential Commission on State Reform was established to consider alternative forms of governance. The 1989 legislation derived from the Commission's report led to the election of governors and mayors by direct vote, and to the allocation of power and responsibility to federal and municipal organisations to control their own development, alongside the National Executive. The legislation transformed the traditionally centralised system, decentralising a degree of autonomous executive and legislative powers to provincial and municipal levels (Estaba, 1991). While political decentralisation for purposes of territorial administration and governance is critical in terms of establishing inclusive democracies within which citizens can shape the development of local and regional spaces, an important mitigating factor has been the role of the private sector. When decentralisation was proposed in Venezuela in the 1970s, the private sector put up obstacles in order to remain in areas of inward investment. It is clear that if decentralisation of political structures does not also carry with it a redistribution of finance and resources, private sector interest groups will be unlikely to respond.

Without significant economic activities in decentralised regions, their power relations with the centre become biased in favour of the centre. In these circumstances it is unlikely that strong regional decentralised identities and activities will develop, establishing what Sergio Boisier terms 'socially constructed spaces', rather than legislative and political constructions in which local citizens construct their own development strategies (Boisier, 1988). A desire to transform political forms of organisation and control and establish alternative political geographies beyond the state characterised Central America during the 1980s. With low intensity conflict a feature of everyday life for most people and with the failure of states to act as resource providers rather than as an aggressive military force, alternatives to the nation-state have emerged. These alternatives are embodied within social movements.

Box 4.2 The local popular governments of El Salvador

Jenny Pearce (1986) reveals one example of the establishment of local alternatives of political organisation, within El Salvador. While state-directed peasant cooperatives and organisations, based on credit assistance, had been established during the 1960s under the US-backed Alliance for Progress, the strength and numbers of cooperatives by the 1980s had created a self-help alternative for many local communities. These groups were easily radicalised by extreme elements of the Catholic Church and revolutionary organisations – the National Resistance (RN) and Popular Liberation Forces (FPL). From these radicalised groups emerged alternative political systems of organisation set up in opposition to the centralised nation-state.

Local popular governments (PPL – *Poder Popular Local*), based on the experience of Christian-based communities and peasant unions, emerged in areas of revolutionary FMLN control from 1982. The goal of the PPLs was popular democracy and political participation, in an attempt to encourage people to stay in their localities in preference to becoming refugees. Each PPL was elected by and responsible to 400–500 people who constituted a 'locality'; this locality was itself comprised of three or four 'bases', smaller units based on traditional *cantones*. In the strongest example of PPL activity, the region of Chalatenango was divided into three sub-zones with each sub-zone operating seven PPLs and a sub-regional government. The PPLs provided an example of the spatial decentralisation of power embodied within the ideals of the FMLN Democratic Revolutionary Government. PPLs encouraged people to stay and be involved rather than migrate.

As with many cases of grassroots organisation, women were well represented on the PPLs. More women had become politicised due to the work of the Association of Salvadorean Women (AMES) established in 1979. This movement encouraged women's participation in political activities following the rise in numbers of female-headed households and orphans, and women's increased role in economic production.

Source: Pearce (1986)

REGIONAL DEVELOPMENT AND RESOURCE FRONTIERS

Regional imbalance

Regions and regionalism have played an important part in the development of the Latin American state and its organisation relative to the national territorial space. In most cases of national development, the region has provided a suitable basic areal unit, in particular for the location of productive activities. Regional development has often preceded territorial delineation, with the construction of regional and provincial boundaries following on, since the extent of a region is most often based on its topographical structure. In many cases river catchments and climatic zones still serve as the basis for the delineation of national territorial regions. The history of these divisions often dates back to the colonial period, as in the case of the donatary-captaincy divisions of Brazil that now underpin the federal state system.

Box 4.3 Worlds apart: Brazil's North East and South East

Among Latin American states, Brazil provides many of the most striking contrasts. The country is at one and the same time an important emerging market which has been fully inserted into the global political economy, and a country which Becker and Egler suggest inaugurated modernised poverty. Apart from Brazil being the most polarised society in terms of income distribution, there are marked regional differences. The North East of the country provides the greatest contrast when compared with the South East. The North East is recognised as a problem region due to the intermittent water shortage crises within the 'drought polygon', while the South East is recognised as the motor of the Brazilian economy based within the geographical triangle of São Paulo–Rio de Janeiro–Belo Horizonte.

Historically, the North East formed the centre of colonial Brazil, based on the plantation economies of the region with their dependence on slave labour. The earliest capital of Brazil was Salvador de Bâhia but the development of the coffee trade further south at the turn of the twentieth century signalled a shift in the 'centring' of the country. The capital shifted with this relocating of the national economy towards the Centre-South. Rio de Janeiro replaced Salvador, before it was itself replaced (1960) by Brasilia, a new capital which was constructed in order to overcome the over-concentration of economic and political activities in the South East of the country.

While the coffee trade marked a geographical shift in the economy away from the North East, the push towards industrialisation also became focused on the South East, building on the agricultural wealth of the Centre-South and South East, and also on the wealth of North East absentee landowners who preferred the quality of life in the urban South East rather than the traditional plantation livelihood.

During the first half of the twentieth century, and especially during Brazil's 'economic miracle' of the 1950s and 1960s, the South East experienced an industrial boom based on the automobile, iron and steel, chemicals and pharmaceuticals industries. But while the South East 'boomed', the North East 'busted' as commodity prices failed to keep pace with manufactures prices. The North East became associated with the colonial past; the South East came to symbolise development, modernisation and progress. A large-scale migration was the social response to economic imbalances that resulted from uneven regional development, North Eastern poverty, seven-year cycles of crippling drought, and the attraction of the industrial urban centres of the South East.

To combat migration to the overpopulated urban centres of the South

---

East, the government initiated two regional development programmes with direct relevance to the North East: the colonisation of Amazonia, and the development assistance of SUDENE (the Superintendency for the Development of the North East). Offers of resettlement in Amazonia for rural migrants from the North East resulted in a rearrangement of the North East migratory pattern increasingly towards the west. However, difficulties in cultivating Amazonian soil, and bureaucratic failures in terms of not reaching resettlement targets and establishing tenure rights, have hindered a solution to the regional problems of the North East. SUDENE was established as a development agency in 1959 to encourage industrialisation and business relocation with attractive taxation policies, and to promote irrigation and other water resource projects in order to curb the excesses of the drought years. Corruption and the existing landholding structures within the region have ensured that SUDENE has been as unsuccessful as the Amazonian colonisation programmes.

The North East still suffers from drought and continued outmigration, and maintains its status as the poorest region in Latin America. In contrast, the economic triangle in the South East continues to act as a magnet for foreign investment, industrial activities and broader Brazilian economic development. Until the North East develops strong movements in civil society to confront the imbalance of institutional politics and policies emanating from the South East, the regional polarisation of Brazil is likely to continue.

Sources: Henshall and Momsen (1974), Morris (1987), Becker and Egler (1992), Pereira (1993)

Regional divisions often lead to difficulties associated with national territorial integration and cohesion, with associated geopolitical anxieties. On its sensitive northern frontier with Peru and Bolivia, Chile's Atacama Desert provides an obstacle to swift transport and communications links. A greater Chilean presence in the northern Atacama in terms of population and state authority (military, police and customs) has been considered geostrategically important for most of the twentieth century, especially the promotion of Arica as a development pole to encourage migration. However, integration through or via the desert is problematic, due to the poor infrastructural systems in the northern regions. Geostrategic goals of integration also serve the purpose of decentralising national populations in order to utilise national resources more effectively. In Chile, 50 per cent of the national population occupies 4.2 per cent of the national space (García Covarrubias, 1985; Von Chrismar Escuti, 1986; Ghisolfo Araya, 1988). This situation is mirrored throughout South and Central America by different regional obstacles that have slowed territorial integration and the construction of an interlinked national space.

While regional development is essentially to be understood in economic terms, its impact transcends economy and integrates political, social, cultural and physical characteristics. Viewing these components as a whole, the idea of territory emerges as the space within which change (development) takes

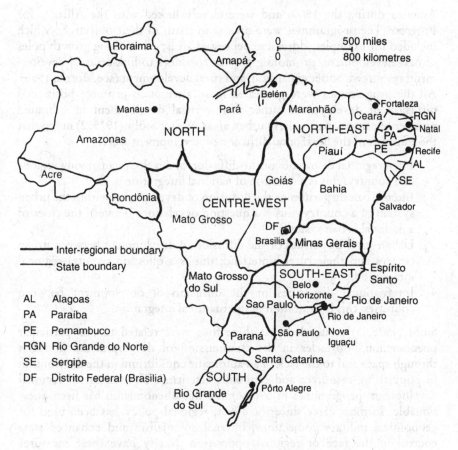

Figure 4.1 Brazil's regional and federal state divisions
Source: Matthews (1984)

place. The organisation of space and its categorisation for purposes of ownership reveals a spatial expression of economic, social, cultural and ecological policies emanating from social groups. For this reason, the conception of space is critical to an understanding of development (or non-development) and cannot be judged by pure neoclassical and structural theories of inter-regional disparities (Fernández, Fernández and Barliza Zubiria, 1991). At a local, regional or intra-regional level, the impacts of planning, organisation and policy-making become more apparent and their socio-economic impacts clearer. On these grounds, it is the importance of spatial policy at multiple scales as much as regional policy, and also the overlap between the two, that is at the root of development and the organisation and control of space.

Regional development programmes increased rapidly throughout Latin

America during the 1960s and were closely linked with the Alliance for Progress. The programmes were diverse in terms of their objectives, which included for example: addressing depressed areas; establishing growth poles of new development; promoting growth corridors to link areas of development; resource frontier expansion; and rural development (see Morris, 1996). All the programmes were generated in an attempt to promote horizontal diffusion of development rather than vertical development in a limited number of centres, particularly urban areas. Walter Stöhr (1975, 2) noted that the concerns of the horizontal diffusion of development were:

1　Inter-regional disparities or the diffusion of development among regions of a country: this is a question of national integration.
2　Inter-urban disparities or the diffusion of development within the urban system of a country: this is a question of relations between the cities of a national urban system.
3　Urban-rural disparities or the diffusion of development between urban centres and their rural hinterland: this is a question of intra-regional integration.
4　International disparities or the diffusion of development between countries: this is a question of international integration.

Stöhr (1975, 3) contended that these four were related to one particular phenomenon, 'obstacles in the transmission of developmental impulses through space and to the lack of an automatic equilibrium in the distribution of population, resources and economic activities'. The success of regional development programmes in overcoming this phenomenon has been questionable. In most cases since the 1960s, regional policy has been used for geopolitical military projections, internal colonialism and enhanced state control in the face of regional opposition. Rarely have these measures achieved greater socio-economic equilibrium (see Slater, 1989a; Lázaro Araujo, 1993). The role of the expansion and decentralisation of capitalism into more peripheral regions of nation-states and the role of the state in using regional policies to counter regional opposition are also important in terms of the failure of regional policies to address issues of socio-economic disequilibrium.

Besides regional policy, territorial policy adopts a wider framework that includes local scales of analysis, and confronts issues of ownership and socio-economic conditions often overlooked by regional policy in its pursuit of geopolitical and economic objectives. Territory not only has a spatio-temporal dimension but has to include the interwoven elements of production, society and the environment (Torres 1993). This also applies to the extension of national space into maritime areas. As productive spaces which often contribute significantly to national production, i.e. the capture fisheries industries of Peru and Chile, and the Venezuelan oil industry, regional development becomes extended beyond its traditional land-based formulations to encompass its 200 nautical miles of territorial sea or Exclusive Economic Zone (EEZ) (Lara, 1991).

In the Andean region, measures have been put in place to introduce greater regional autonomy. The 1989 restructuring of Peruvian territorial administration, a movement away from the centralism that has hampered regional economic cycles, has given rise to the promotion of Regional Development Corporations for the purpose of decentralising public administration. For the most part these organisations have been unsuccessful, due to their uneasy situation between community structures and central government. Central government has been especially restrictive, because of its political tradition and representation of sectoral interests (Gierhake, 1993). Within the corporations, limited participation and weak policies prevented further financial responsibilities and extended influence. This Peruvian experience is similar to many throughout Latin America in that there is ongoing dispute as to who is responsible for regional development, whether it is top-down or bottom-up, and what functions and resources should come under regional auspices in terms of spatial planning, environmental issues and socio-economic activities.

The case of post-revolutionary Nicaragua provides a significant contrast to the republics still dominated by landed interests. A 1987 report of the Nicaraguan Institute for Territorial Studies created a National Reference Plan for Physical Planning. This document provided the framework for regional organisation within the country. By establishing regional factors of natural resources, production, distribution and consumption, settlements, levels of agricultural development, levels of industrial development of urban centres, and the road networks, a national (re)organisation of territorial space could be carried out. The intention of the plan was to obtain greater food security by making changes to land use, away from the traditional agroexport model (Leonelli and Benito, 1991). Despite the limitations of such a large undertaking, the plan assisted in establishing an integrated vision of the national territory and a reference point for further regional and sectoral planning.

The role of regions in terms of the construction of the nation-state 'whole' is a critical one. Rudolf Kjellén established the concept of the 'organic' state in the late nineteenth century, suggesting that states are not fixed in time and space, that they evolve and expand or contract according to the territorial and resource needs of societies within those states. Due to the relative youth of the states of Latin America, there are still many unresolved disputes of considerable historical weight resulting from territorial organisation dating back to Iberian colonialism and the period following independence. This is the case on the inter-state level and is embodied in the militaristic geopolitical tradition and in the notion of patriotic territory that is advanced in public education.

Resource frontiers

On the intra-state level there are the territorial issues that precede the formal inter-state system, in terms of land ownership and rights of indigenous collectivities, in southern Chile for example, also Amazonia and Central America. These inter- and intra-state territorial issues are critical, because of the significance of power-space relations at all levels, from the shanty towns to ministries of foreign affairs. Resource control and management is an integral element of this territoriality. It is for this reason that Amazonia and Antarctica have become symbols of inter-state territoriality within the continent.[5] Both have been identified as areas of geopolitical significance based on potential resource wealth. However, the status of these two frontier regions is different. The region of Amazonia is Brazilian sovereign territory for the most part, while Antarctica is covered by international agreement as a shared resource. It is this fundamental difference of sovereign territorial control in the one instance and international agreement in the other that reveals the essential problems associated with areas constructed as frontier or resource areas.

The notion of internal colonialism, or internal imperialism, is strong in Latin America (Hechter, 1975; Cano, 1990). With vast territorial spaces still largely unsettled for reasons of low productivity or harsh environments, especially in Brazil and Argentina, successive state administrations have established strategies to encourage migration to and settlement in remoter areas of the national space for geopolitical reasons (Hepple, 1986; Morris, 1996). The process has been one of expansion to fill 'the hollow continent' as Bromley and Bromley (1988) describe it. The remote areas have been constructed as frontier areas where opportunities abound for land ownership and self-employment and population movement has been facilitated by new transport and communications networks (Saint Lary and Colantuono, 1993). These migrants are Latin America's late twentieth-century pioneers. The importance of frontier areas for expansion has been central to the notion of the development of the nation-state since independence. Alistair Hennessy (1978) highlights the wide range of types of frontier area that have been developed within the continent: mission; Indian; maroon (of runaway slaves); mining (gold, silver, base metals); cattle; agricultural; rubber; Anglo-Hispanic (as on the Mexican-US border); and political. It is this wide range of factors lying behind expansion that gave rise to the gradual integration of the continental space into state territories during the nineteenth and twentieth centuries.

The existence of pioneers and frontier regions performs several functions. First, there is the geopolitical imperative. Second, these regions act as a safety valve for pressures on the nation's core regions. Third, there is the potential resource extraction and production opportunity (Coy, 1992). The Amazon provides the best example of such a frontier region with a strong pioneering identity. It is Brazil's 'Wild West' and since its 'opening up' in the early 1960s it has distracted social pressure from the burgeoning South-east of the country and the drought polygon in the North-east.

Within Latin America, for reasons of environmental management, ecology, indigenous land rights and Brazilian economic growth, Amazonia has been a focal point for national development and international interest and investment. From the early 1960s, Amazonia was targeted as a region for development by the authoritarian regime since it comprises half of the national territory. The 'march to the West' that accompanied Brazil's authoritarian development strategies marked a turning point for the Amazonia region (Foweraker, 1981; Hepple, 1986). The development strategies included economic activity and settlement, referred to as 'colonisation', assisted by state infrastructural programmes. The goal was for pioneers to tame the wild frontier, clearing the way for the expansion of Brazilian capitalism.

The construction of Amazonia as a frontier region of colonisation and development provided a suitable distraction from the problems experienced within the states of the North-east and the overwhelming urbanisation patterns of the South-east. For those prepared to resettle in Amazonia, they were constructed in the popular consciousness as pioneers. For the state, the pioneer towns, either 'company towns' for mining operations or spontaneous settlements of peasant migration, became an extension of control over the national territory. Coy (1992, 10) writes that 'the pioneer town is first and foremost the place of government presence and state control on the pioneer front, equal in importance to the societal control over the peripheral development by the centre'. The continued profitable extraction of minerals and forest products perpetuates the myth of unlimited opportunities in the pioneer front which feed back into the centre (Volbeda, 1986). With such a notion of Amazonia as an 'internal colony', of not only Brazil but other neighbours, as a spatial escape valve, projects suitable to the region cannot be implemented (Kohlhepp, 1989).

It is not only Brazil that has interests in extending into the Amazon basin. Ecuador also sought to exploit its Oriente province resources under President Febres by encouraging multinational production and extraction within the area. The oil companies have constructed the infrastructure by which the interior areas became accessible from the 1960s. This has had a negative effect on indigenous groups in the area in terms of their populations as their lands have been opened up for economic exploitation and they have been exposed to evangelisation (Corkill and Cubitt, 1988; Radcliffe and Westwood, 1996). As in the Brazilian case, the movement of production to the region has been driven by natural resource exploitation, to reduce land use pressure elsewhere in the country, and to act as a strategy for national territorial integrity.

The Peruvian Amazon, as with other republics bordering the region, is also constructed as an area to be exploited for national territorial purposes and resource exploitation. President Belaúnde noted that the future of Peru lay in the jungle as he inaugurated the Peruvian Amazon urban centre of Constitución (Reid, 1984). The obstacles that have been faced, regardless of nationality, have been consistent: impoverished soils unsuitable for agricultural production; land rights issues relating to indigenous groups; the expense of prestige projects to establish a competent infrastructure for transport and communications; and uncontrolled colonisation and economic activities.

Contrasting with Amazonia, the Antarctic continent, by its proximity to the Latin American continent, has also provided another focal point of Latin American geopolitical attention. Among the many geopolitical issues within the continent, the Antarctic provides a particularly complex problem of multiple state interests, suggesting that the frozen continent will be central to Latin American geopolitical debates into the next century (Child, 1990). Unlike the Amazon, which is located within national spaces (principally Brazil), the Antarctic has the status of a global commons, a shared common space, in that the international 'community' has reached agreement on its non-development and its scientific value.[6] This contrasts strongly with the use of other global commons, particularly fisheries which are in crisis having been overfished in common waters.

Rather than being an economic frontier, the Antarctic continent has been constructed by the international community as a 'frontier of humankind', in the same way as outer space and the ocean floor have been. The 1959 Washington Treaty on the Antarctic, signed by twelve states (seven of which had territorial claims), defined the region as a demilitarised zone, as a scientific laboratory and an ecological reserve, thus establishing an Antarctic Treaty System (ATS) (Zegers Santa Cruz, 1983; García Monge, 1990–91).[7] In this way, any national territorial rights to the region in terms of sovereignty have been suspended.

To some countries this situation does not fit well with national geopolitical rhetoric. On a visit to the Chilean base of Villa las Estrellas in Antarctica, General Pinochet declared (*Boletín Antártico Chileno*, 1984, 59–61): 'You, compatriots, with the strength of spirit of pioneers are sending a light into the future from the present, which honourably represents the intention of the government and the Chilean people to develop this region.' Daniel Contreras, the first mayor of an Antarctic community (signifying more than a scientific location in government terms), replied by announcing that: 'Our presence here is the best guarantee of Chilean rights in this territory and of its permanent interest in keeping our sovereignty.'

Until the Washington Treaty, the continent was an area of geopolitical uncertainty. In 1939–41, taking advantage of distractions in Europe, a US expedition to Antarctica constructed bases for US sovereignty claims. Chile reciprocated in 1947 by creating a zone of reciprocal assistance around the continent. Chilean geopolitics maintains a strong focus on Antarctic affairs

(Gajardo Villarroel, 1983; Merino Castro, 1987);[8] Bustamente (1988, 3) describes the Antarctic as the southern 'El Dorado'. The Cold War also made clear the importance of the Antarctic in terms of geostrategy relating to control over the South Pacific Ocean, and vice versa (Riesco, 1985). Both US and Soviet submarines were active in this area. The post-Cold War era has seen the intensification of scientific research. However, Antarctica's significance for many countries as a site for potential resource development poses the question, how long can the nearly forty-five-year-old Washington Treaty continue to keep all its signatories united in a common purpose for the collective utilisation of one of the last remaining global commons?

Amazonia and Antarctica provide contrasting examples of complex resource and territorial issues. Rights over resources, latent or potential, underlie much of the geostrategic manoeuvring within Latin America. The geostrategic arguments that underpin the economic ends of resource exploitation and the development of the national space, or perceived national space, have been carefully constructed. Brazil's 'opening up' of the Amazon, Chile's territorial claims over Antarctica, Argentina's claims over the Falklands, Ecuador and the Galapagos: all these issues, whether internal to the nation-state, acting within sovereign spaces, or external to the state in terms of historical claims and territorial 'rights' over shared or 'other' sovereign space, may lead to conflict. More often than not there is the promotion of a pioneering consciousness to encourage population movements into disputed areas, or to engage in productive activities.

Beneath the rhetoric of geopolitics, the existence of economically active populations, as opposed to an overt military presence, within 'frontier' regions is a powerful weapon in negotiations and claims. It is for this reason that spaces unsuitable for production and settlement have been internally or externally (outside sovereign territory) colonised. The promotion of the concept of frontiers and pioneers ensures that the pursuit of 'something better' for many people in Latin America is maintained and that the nation-state achieves its geostrategic objectives and a territorially unifying national project of integration.

MIGRATION, URBANISATION AND URBAN SPACES

Pre-Columbian and colonial urban space

While regional development provides the basis for most national development strategies, the internal space of the region is highly complex. The spectrum of rural to urban provides a wide variety of socio-economic conditions, obstacles, needs and power-space geometries. In Latin America, regional spaces have undergone rapid change during the twentieth century in response to changes in economic production strategies and spatial policies. The most significant of these changes has been the persistent failure of governments, particularly authoritarian regimes, to engage with the power relations within rural areas. This has led to outmigration in response to agrarian changes such as mechanisation and the introduction of different crops and livestock.

The urban centres of the continent have been the recipients of the migrant flows and have also undergone economic changes, most importantly through the processes of industrialisation and its associated socio-economic multipliers, such as the rise of a service sector, attracting former rural inhabitants. Due to their rapid increases in population from inward migration and natural growth rates, the largest cities have become sprawling industrial complexes that are spatially and socially fragmented. These urban areas are highly complex spaces that have to be analysed by establishing their political geographies and the power relations between and within urban communities in terms of spatial organisation (see Roberts, 1978, 1995).

The organisation of urban space in Latin America precedes the empires of the Incas, Aztecs and Mayans. Cuzco, for example, the Inca city that was the core of the 'empire of the sun' reveals a pre-Incan urban morphology. Predating the Inca urban model in the region was that of the Aymara. Within Aymara society, the Collana comprised the hegemonic ethnic group and would continue to dominate later within the Inca system. The Aymara utilised a simple dualistic social and spatial division of urban areas based on *urcosuyo* and *umasuyo*. The *urco* was associated with masculinity, force and social order while the *uma* was related to fertility and femininity. This division was not sexual, however. The *urco* group was of highland origin and the *uma* of lowland origin but it was the *urco* which held the central urban space (Savard, 1992). As such it is clear that in pre-Inca and Inca urban organisation, divisions of space were also social divisions that fitted into an urban social hierarchy dictating the redistribution of resources.

The spatial organisation of the early urbanism of the Inca empire was also based on ethnicity. Manco Capac, the first leader of the Inca dynasty, put an end to ethnic rivalry and Cuzco became a multi-ethnic centre. However, it still retained urban ethnic divisions of centre and periphery. The Inca urban model worked over the ancient system of *ayllu* and attempted to unify them in order to create a united multi-ethnic urban space. Under the Incas, the sacred centre of Cuzco was dominated by the Collana (the Incan conquerors), the Payan between centre and periphery, and the Cayao on the periphery (Wachtel, 1971; Savard, 1992). The city was effectively divided into four areas (Collana, Cayao, Payan, Collana/Cayao), representing the symbolic significance of the four 'ages' of thousands of years each, also the four parts of the world: between the Pacific and the west, and the Amazon forest and the east, between Quito in the north and Valparaíso in the south. As a result of this urban organisation, the spatial order reflected a social order and wider historical and geographical beliefs. Apart from a limited number of societies, which were organised around a strong hierarchy of theocratic nobility and servile classes, the majority of pre-Columbian societies were essentially nomadic cultures without urban centres, peopled in dispersed groups.

The Inca empire's control of space was founded upon strong central-isation and a social hierarchy (see Katz, 1978). Totalitarianism and state collectivism characterised social organisation, while imperial domination was maintained via the tools of economic tribute, control of language and ideology. All of these features were subsumed within a theocratic system (Muscar Renasayag, 1989). Cuzco was the centre of the system, from which governance emanated and to which wealth was delivered. A simple core-periphery model was established and a road infrastructure maintained the threat of rapid reaction to imperial dissent as well as an effective communica-tions and trade network. Pinto Serrano (1986) notes a similar urban structure within Chibcha society, which was organised along strict religious lines around which civil obedience revolved, with an urban hierarchy based on urban-regional spatial units supporting higher tiers of the urban structure via a tribute system. These models were continued and adapted by the *conquistadores* that followed.

A more rapid development in structure and number of cities in the Americas came about during the Iberian colonial period. As structures for controlling space and extending authority over neighbouring territories, cities became the centres of the 'Europeanisation' of the continent. They existed as spaces independent of indigenous groups and cultures, preserving a Christian-western urban identity as an extension of the imperial centre of Spain (Muscar Renasayag, 1989). Each city drew on the resources of the surrounding area and on a wider scale a system of nodal points, or hispanic nuclei, was established over time to develop a network of economic, administrative and communica-tions routes in order to extend influence over a wider area and to maintain unity across regional divides within the continent.

The experience of Brazil was different from that of the Spanish empire in that, until the 1800s, the country was still predominantly rural in its social and cultural structuring. Brazilian space was organised for plantation production focused on the north-east of the country. Only in later centuries would the centre of attention switch further south. The sheer size of Brazil and its territory reduced the rate at which the Portuguese urban centres constructed the urban networks that had been established in Spanish Latin America by the seventeenth century. Rather than colonial integration, Portuguese centres were oriented towards the sea and maritime trade rather than towards terrestrial connections.

From the sixteenth century, Latin America became a projection of the European mercantilist and bourgeois world. The urban spaces provided the geographical locations for this 'New World' and wealth from surrounding territories flowed to them. The territorial jurisdiction of the city was very much dictated by the area from which it drew its resources, often superimposing upon the existing pre-Columbian spatial and urban structures.

Muscar Renasayag (1989) identifies five types of colonial cities according to function:

1   Representative city: political, governmental, religious or administrative head.
2   Port city: head or terminal of oceanic routes or between regions.
3   Cities of economic development: livestock; agricultural; artisanal.
4   Cities with a commercial function: such as port cities, or in the interior, as key points on road networks.
5   Cities with a strategic function: for the advance or defence of conquered territory.

Colonial cities had varying levels of reciprocity and interdependence due to factors of accessibility and communication. In terms of internal organisation, the development of the *cabildo* system of mayors for city governance enabled a political structuring of urban spaces. A considerable element of the urban function at this time was the attempt to ensure supplies of goods from surrounding areas and the exchange of surplus with other centres. Essentially, urban centres centralised administrative and service functions for surrounding production areas. Consequently, power became centralised within these spaces.

In terms of the planning of the colonial cities, the Law of the Indies, established in the sixteenth century, laid down a uniform planning model. Under the Law, urban spaces became organised around a central *plaza*, around which the institutions of crown, church and civil authorities were located. Also, around the central *plaza*, a grid system encompassed the residential areas of the recognised social groups such as Europeans, creoles and *mestizos*. However, the more peripheral areas, where indigenous groups were located, remained unplanned. This situation set up a historical precedent for the organisation of cities into the late twentieth century, whereby city plans, zoning patterns and planning regulations fail to accommodate the wider urban population (Angotti, 1995).

Post-independence cities

Following independence, cities retained the colonial style of urban spatial organisation. However, the influence of modernism and European urban planning during the late nineteenth century led to a significant redevelopment of many of the largest cities during the early twentieth century. The 1910 remodelling of Mexico City and Buenos Aires, both celebrating their one hundred years of independence (*centenario*), embodied these modernist influences emanating from Haussman's avenues, élite residential areas and public spaces developed in Paris, and from the garden cities of Britain and the USA (Tenorio Trillo, 1996). The basis of the development was a nationalist project for a modern, international and cosmopolitan city termed the 'ideal city'. As with traditional planning, indigenous groups were forced further out to the margins of the city in the redevelopment process. However, a month after the Mexican *centenario* was celebrated, the Mexican revolution led to the social restructuring of the city and its later reorganisation during the 1940s. This later reorganisation represented the power of the PRI in contrast to the former city structure which was a remnant of the Porfirian élites of the turn of the century.[9]

With independence, urban centres still retained their growth rates. From 1850 these rates increased as a response to changing economic activities in rural areas and the development of urban centres. In Colombia, three phases of urban development can be traced: colonisation until 1850, characterised by clear Spanish-Indian demarcation and urban centres as a nucleus of a surrounding region; 1850–1930, during which urban centres expanded considerably in response to disadvantageous conditions in rural areas; 1950 onwards, which was marked by rural violence and an exodus to urban centres (Pinto Serrano, 1986). While each republic and each urban centre has been subjected to different urbanisation rates, urban areas have grown dramatically during the later twentieth century from a slow expansion during the late nineteenth and early twentieth century. Industrialisation has been the motor behind this process (see Butterworth and Chance, 1981; Gwynne, 1985).

Urban centres have provided a focus for struggles of power and space, since half of Latin America's population now lives in cities of over 100,000 people; Angotti (1995) expects this figure to rise to half in cities of over 1 million by the year 2000. With many rural areas still locked into semi-feudal systems of landownership, it is the increasing pressures within urban spaces that have concentrated opposition and popular activity (Castells, 1977; Comacho, 1993). These struggles have developed from the opposition to state planning authorities, private sector landownership, and the pressures applied by the urban poor to establish settlements and the installation of services, such as electricity, water and sanitation (Gilbert 1992, 1994).[10] With weak urban public sector capacity to accommodate the needs of burgeoning urban populations, resulting from ineffectual taxation regimes and national indebtedness, the urban poor have relied on self-help mechanisms in order to provide basic needs.[11]

The late twentieth century has seen the shift of poverty to urban rather than rural areas as a result of high levels of migration. Bryan Roberts (1978) coined the phrase 'cities of peasants' to describe this phenomenon. The process has brought with it a shift in popular politics, with an increasing emphasis on urban problems rather than the condition of rural areas. For the

Table 4.1 Urban population (% of total)

Country	1970	1980	1990	% in main metropolitan area
Argentina	78.4	82.7	86.2	33.4 (Greater Buenos Aires, 1990)
Bolivia	40.7	44.3	51.4	13.8 (La Paz, 1980)
Brazil	55.8	67.5	76.9	10.4 (São Paulo, 1990)
Chile	75.2	81.1	83.6	35.0 (Santiago, 1990)
Colombia	57.2	64.2	70.3	14.8 (Bogotá, 1980)
Costa Rica	39.7	46.0	53.6	25.1 (San José, 1980)
Cuba	60.2	68.1	74.9	19.8 (Havana, 1980)
Ecuador	40.0	47.3	56.9	15.5 (Guayaquil, 1980)
El Salvador	39.4	41.5	44.4	13.9 (San Salvador, 1970)
Guatemala	35.7	38.5	42.0	19.9 (Guatemala City, 1980)
Haiti	19.8	24.6	30.3	14.3 (Port-au-Prince, 1980)
Honduras	28.9	35.9	43.6	10.3 (Tegucigalpa, 1980)
Mexico	59.0	66.4	72.6	18.5 (Mexico City, 1990)
Nicaragua	47.0	53.4	59.8	21.2 (Managua, 1970)
Panama	47.6	50.6	54.8	33.3 (Panama City, 1980)
Paraguay	37.1	41.7	47.5	38.5 (Asunción, 1980)
Peru	57.4	64.5	70.2	29.0 (Lima, 1990)
Dom. Rep.	40.3	50.5	60.4	23.3 (Santo Domingo, 1980)
Uruguay	82.1	83.8	85.5	49.5 (Montevideo, 1980)
Venezuela	72.4	83.3	90.5	16.5 (Caracas, 1990)

Source: CEPAL, Statistical Yearbook for Latin America and the Caribbean, 1994

same reason, there has been a shift in the political geographies of nation-states, with urban areas receiving greater attention as the centres of opposition and change for the nation's lower income groups.

LAND TENURE AND REFORM

Landholding structures

Behind the issues of urbanisation, economic production and orientation, and levels of socio-economic development, lies the landholding structure and appeals for its reform. The historical context of land ownership since the Iberian conquest has been one of concentration in a few hands. Throughout the colonial period and into the post-independence era, rural production was characterised by small landowning élites and a large tied agricultural labour force, working as tenants on estates (haciendas and latifundios) on a subsistence basis, or in exchange for their most basic needs. It is these rural social systems that gave rise to the prevalence of the campesinos, or peasants, in rural areas living by subsistence means and also selling their labour.[12] These livelihoods in turn gave rise to high levels of rural poverty and the 'push' towards urbanisation.

The predominance of these types of system have earned the agrarian

Plate 4.5 The demographic explosion of Latin American cities has led to inadequate housing and infrastructure for the majority of the urban population. Asunción (Paraguay).

productive structures the title semi-feudal and semi-capitalist: feudal in social structure, capitalist in terms of the sale of production (Roxborough, 1979). Appeals for the reform of these persisting agricultural systems and land-holding patterns have given rise to considerable social unrest and to insurrection throughout the continent during the twentieth century. Only by

Plate 4.6 Recent arrivals to the cities often construct makeshift dwellings until further materials can be collected for the consolidation of the structure. Lima (Peru). Photo: Nina Laurie.

Table 4.2 Changes in the extent of poverty, 1970–90

| | Poor population[a] (%) | | |
	Total	Urban	Rural
1970	119,800 (45)	44,200 (29)	75,600 (67)
1980	135,900 (41)	62,900 (30)	73,000 (60)
1986	170,200 (43)	94,400 (36)	75,800 (60)
1990[c]	195,900 (46)	115,500 (39)	80,400 (61)
	Indigent population[b] (%)		
	Total	Urban	Rural
1970	63,700 (24)	19,900 (13)	43,800 (40)
1980	62,400 (19)	22,500 (11)	39,900 (33)
1986	81,400 (21)	35,800 (14)	45,600 (36)
1990[c]	93,500 (22)	44,900 (15)	48,600 (37)

[a] People having incomes below the poverty line. Includes people living in indigence.
[b] People having incomes below the indigence line, i.e. living in extreme poverty.
[c] Estimates for 19 countries in the region.
Source: ECLAC, *Social Panorama, 1994*

Table 4.3 Open urban unemployment, selected countries (first half of each year)

	1991	1992	1993	1994	1995[a]
Argentina[b]	6.9	6.9	9.9	10.8	18.6
Brazil[c]	5.5	6.0	6.0	5.4	4.4
Chile[d]	6.5	5.0	4.2	6.0	5.7
Colombia[e]	10.5	11.0	9.4	10.1	8.6
Ecuador[f]	—	—	9.4	8.4	—
Mexico[g]	2.5	2.9	3.3	3.6	5.8
Uruguay[h]	9.7	9.7	8.9	8.5	10.5
Venezuela[i]	10.3	8.4	6.9	8.9	11.0

[a]Preliminary estimates. [b]Urban nationwide (data from May each year). [c]Six metropolitan areas. [d]Santiago Metropolitan Region. [e]Seven metropolitan areas. [f]Urban nationwide (July each year). [g]Urban nationwide. [h]Montevideo. [i]Nationwide total.
Source: ECLAC, *Economic Panorama of Latin America, 1995*

incorporating *campesinos* and other rural groups into rural development and agricultural programmes can rural poverty be confronted and the continued migration to urban areas slowed.[13] While reforms have been slow to emerge within most political systems, post-revolutionary or socialist political regimes have recognised the fundamental importance of land reform in terms of social change, as in Cuba, Nicaragua, Chile (1970–73), Peru (1968–75) and Mexico.

The process of reform

The Mexican land reforms of the early twentieth century, the system of community land provisions known as *ejidos*, were significant in their timing and scale. However, the legacy of the reforms lives on. The late twentieth-century 'rebellion' in Chiapas is as much a response to the failure to initiate land reform in Chiapas state in the 1920s and 1930s as it is a resistance to the same party that initiated the reforms and remains in power, the PRI. It is also a reaction to the formal end of land reform embodied in the 1992 Article 27 amendment to the constitution. The Article provides the legal basis for privatising the *ejidos* and is in the context of the liberalisation of agricultural trade that began in 1990.

Peasant movements have responded to the increased liberalisation of the agricultural sector by pressurising the government over issues such as crop prices, credit, debt renegotiation and input subsidies. These groups are linked within Mexico under the umbrellas of the Plan of Ayala National Network (CNPA) and the National Union of Regional Campesino Organisations (UNORCA). However, despite this type of organisational activity and pressure, even the conservative estimate of a World Bank study estimates that the impact of agricultural liberalisation policies will result in a rural out-

Table 4.4 Major agrarian reforms in Latin America

Mexico	1917	Division of *haciendas*; half of all farm land reformed into communal tenure with individual rights.
Guatemala	1952–54	Idle land and rented land transferred from large estates to small farmers and cooperatives; reform reversed after 1954 coup.
Bolivia	1953–70	Division of *haciendas*; much of the cultivated land transferred to tenants.
Cuba	1959–63	Initial expropriation of idle land, and land transferred to tenants; all large farms later converted to state farms.
Venezuela	1959–70	Some division of *haciendas*, much settlement of newly irrigated land.
Chile	1967–73	Division of *haciendas*; cooperatives and, after 1970, collectives encouraged; reform reversed after 1973 coup.
Peru	1968–76	Division of *haciendas*, expropriation of plantations; promotion of cooperatives.
Ecuador	1973	Some division of *haciendas*, land to tenants especially on the coast; considerable sale of land.
Nicaragua	1979	State farms on the lands of the ex-dictator, expropriation of idle land, land transferred to tenants, promotion of cooperatives.
El Salvador	1980	Division of very large estates, some land to small-scale tenants.

Source: Adapted from Townsend (1987)

migration of 1.1 million workers over ten years, increasing to 1.9 million if maize production is also fully liberalised (Foley, 1995).

Box 4.4 The *Zapatistas* of Chiapas: resistance for reform

On New Year's day 1994, the Zapatista National Liberation Army (EZLN) initiated an armed insurrection that resulted in the loss of hundreds of lives during the conflict with the army in the following two weeks. It was not until March 1994 that a first round of talks was set up after a long military stalemate within and around Chiapas state's jungle areas. A few days after the talks began, the PRI presidential candidate Luis Colosio was assassinated in Tijuana. In September 1994, the Secretary General of the PRI, José Ruiz Massieu, was also shot dead. However, both murders were organised from within the ranks of the PRI and not from Chiapas. Immediately Sub-comandante Marcos, the leader of the EZLN, and the insurrectionists faced a different national context, of fracturing within the party-state system of the PRI.

The EZLN is not a Marxist-Leninist organisation, like Sendero in Peru, committed to armed struggle and the take-over of state. It is a popular resistance movement that has taken up arms with radical democratic motives rather than ideological ones. The 70,000 in the ranks of the Zapatista army are peasants, principally Mayans. They consider themselves to be the victims of the PRI's failure to initiate agrarian reform and enduring policies of exclusionary racism. The EZLN mobilised itself in 1994 after a long campaign that began on a small scale in 1983 following an increase in uncoordinated peasant struggles since 1974.

In 1983, Chiapan peasants were suffering in the face of increased numbers of Guatemalan migrants and refugees (c. 80,000) who were employed cheaply in coffee production, also with respect to the long-standing land tenure inequalities in the state, a legacy of the failure to pursue land reform within the region. The events of 1993 proved to be the last straw in terms of the choice between popular resistance and armed resistance, ending ten years of peaceful political opposition. The Article 27 amendment of the constitution pushed through by President Salinas established rural land markets and the cancellation of state-sponsored redistribution of land. It also reduced access to previously common woodland and introduced a new penal code that criminalised opposition to these new measures. Article 27 will compound the concentration of land ownership: currently 0.16 per cent of coffee farmers work 12 per cent of coffee production land in Chiapas. The Zapatistas responded by mobilising themselves against Salinas and the forces of the PRI.

The Zapatistas' aims are clear. They desire free elections, the expansion of peaceful, popular political participation, and the establishment of alternative forms of government at local, municipal and regional levels to ensure agrarian reform is passed and implemented. Although talks continue to be held, an agreement would appear to be a long way off. Nevertheless, the Chiapas mobilisation has been held up as a model of civil opposition to landholding inequalities.

Sources: García de León (1992), Hernández (1994)

All countries have varying degrees of landholding inequalities, but Brazil reveals the largest imbalance in terms of numbers of landowners and the size of their holdings. In 1985, when the military loosened their grip on power, the appeal was for land reform but it was soon clear that rather than land reform, the measures put in place were oriented towards land reformulation. All regional questions relating to Brazil fall into the debate on land reform. Brazilian landownership falls into the theoretical realm of internal colonialism or internal imperialism (Hechter, 1975; Cano, 1990). As, during the twentieth century, the national 'centre' of Brazil has become situated within the triangle of São Paulo, Rio de Janeiro and Belo Horizonte, inter-regional flows of trade and investment have led to the marginalisation of regions still controlled by landowning élites and old commercial capital.

Only through a strong regional policy with effective land reforms, rather than through the centralising effects of domestic economic policies, can socio-economic conditions be stimulated within Brazil's enormous periphery. Policies directed at improving land tenure and settlement on productive land in the northern regions of the country must be the centrepoint of national development strategies. The economic deceleration experienced during the 1980s has had marked effects on urban as well as rural areas, therefore there has been increased attention to discouraging migration to the largest urban areas, and attempting to resettle potential migrants away from the magnet of the south-eastern states.

Much of the opposition to land reform is to the fragmentation of land and the impact that this will have on agricultural production, and on the landowning élites. In Chile in 1979 a new piece of legislation decreed that all Mapuche Indians divide their lands and become small landholders. This was an attack on the communal landholding system. Since 1492, communal lands have been seen as an obstacle to national economic objectives. They are seen as inefficient and under-capitalised, obscuring the functional purposes of subsistence and locally organised trading networks (Galeano, 1991).[14] Recognition that traditional power structures have been generated by association with agricultural production and systems of labour exploitation reveals the difficulties linked with land reform. The redistribution of land is not only the physical fragmentation of territorial space but is also an attack on the social systems that have been developed within the power-space relations embedded within rural organisation since the colonial period. To change the organisation of space, therefore, is to change power relations over space.

FURTHER READING

Overviews of Latin American politics, including regime types and elections, may be found in Gary Wynia's *The Politics of Latin American Development* (1990), Jean Grugel's *Politics and Development in the Caribbean Basin* (1995) and James Dunkerley's *The Pacification of Central America* (1994). Walter Stöhr's *Regional Development: Experiences and Prospects in Latin America* (1975), Arthur Morris's *Latin America: Economic Development and Regional Differentiation* (1981), Morris and Lowder's *Decentralization in Latin America* (1992) and Bob Gwynne's *Industrialisation and Urbanisation in Latin America* (1985) account for the regional development policies of the 1960s to 1980s, while agricultural change and the urbanisation process are covered in Bryan Roberts's *The Making of Citizens: Cities of Peasants Revisited* (1995) and Alan Gilbert's *The Latin American City* (1994).

5

COMMUNITIES OF
POVERTY, BODIES OF
POWER

From squatters to ecologists, from popular kitchens in poor
urban neighbourhoods to Socialist feminist groups, from human
rights and defense of life mobilisations to gay and lesbian
coalitions, the spectrum of Latin American collective action
covers a broad range. It includes, as well, the movements of black
and indigenous peoples; new modalities of workers' cooperatives
and peasant struggles; middle- and lower-middle-class civic
movements; the defense of the rain forest; and even cultural
manifestations embodied, for instance, in Afro-Caribbean musi-
cal forms (such as salsa and reggae) and incipient antinuclear
protest in some countries. This rich mosaic of identities is at the
heart of our project. Clearly, it represents a changed social,
cultural, economic and political reality in the continent.

(Escobar and Alvarez 1992, 2)

MULTIPLE IDENTITIES AND SPATIAL POLITICS

Gendered identities

When pre-Columbian societies came face to face with the European
Renaissance societies of Spain and Portugal, a long history of cultural
relations was set in motion. These cultural relations have been expounded in
poetry and literature, realised through politics and economics, and grounded
in space. This primarily racial encounter has been extended into other diverse
encounters that have been established and resisted within the Latin American
continent during the subsequent 500 years. These include encounters of
gender, race, ethnicity, sexuality and age (see Radcliffe and Westwood,
1993).[1] Together, they comprise the complex web of connections that have
helped to construct group identities throughout the continent. The spatial
organisation of these groups and their activities contributes to the constructions of Latin American political geographies in terms of the power and space

155

relations within and between distinct social groups. These relations of culture and power within Latin Américan society can be highlighted at the expense of the traditional focus on state institutional power, as Daniel Levine (1993, 14) suggests:

> The role of institutions is often painted in mostly negative terms, with attention to repression, exile, purges, and censorship. These are important but by themselves fail to capture the full range and dynamic of culture and power.

A key identity issue has been that of the role and representation of women and men within Latin America. Women's movements and groups have played a central part in many revolutionary activities, grassroots projects and economic reorganisation. To write about social change, economic change and political change requires a conceptualisation of gender roles and activities. Many Latin American women have broken or resisted traditional constructions of womanhood, transgressing the patriarchal structuring of societies to establish new spaces of identity within which they can act.[2] The most important social construction at the heart of the gender debate is that of *machismo* and its reverse, *marianismo*.

The Latin American stereotype of masculinity is that of *machismo*. Within this term can be found the basis for a particular construction of maleness: as dominant, the breadwinner, the family head and the decision-maker. The woman, as the counterpoise to the *macho*, is constructed as the carer, the reproducer and home-maker. This collection of constructed cultural roles is referred to as *marianismo*, an extension of the virtues of the Virgin Mary (Maria). This leads to the simplistic binary choice constructed for women: the virtues of the Madonna or her antithesis, the whore.

The *macho* society is the brand of Latin American patriarchy that still predominates within Latin American societies. However, resistance to that type of social structuring is increasingly making its voice heard and its strength felt. Clearly the historical experience of this type of society has established rigid boundaries within which women, and non-*macho* males, have struggled to assert themselves. A gendered power relationship is determined by the patriarchal system. Men should be dominant, strong and in control of women and children. Women should be subservient and supportive. Increasingly women are challenging these stereotypes. However, the traditional model of society remains strong and continues to be central to the structuring of Latin American spaces, from the household to national territorial spaces.

The gendered power relationships that evolve from the predominant social stereotyping of the patriarchal system have powerful impacts in space. The most obvious is the construction of domestic, or private, spaces and non-domestic, or public, spaces. Within patriarchal constructions of space and society, women are associated with and considered with regard to private spaces while men are linked with public spaces. This holds with the basis of the patriarchal society in which men are considered predominant and more powerful, therefore they dominate within society's shared public spaces. Public spaces are organised and controlled by *macho* norms. The private domestic space is reserved for the norms of *marianismo*; it is a place for reproductive activities such as caring. While the challenges to the patriarchal structure within public spaces emerge within debates on women's labour and economic roles, and security (from rape for instance), within private spaces it is domestic violence (which may also include rape) and the sharing of household and reproductive activities that receive greatest attention.[3]

During the 1970s and 1980s, women became more active in public spaces within social movements. Of particular significance was their role in self-help and human rights organisations. Lessie Jo Frazier and Joesph Scarpaci (1993) write about the gendering of landscapes and the contestation of state terror in terms of the pivotal roles of women in public protest and opposition. In Argentina, the Madres de la Plaza de Mayo (Mothers of the Plaza de Mayo) continue to protest every Thursday at the lack of information about, and justice for, Argentina's *desaparecidos*; their protest began in 1977 (Radcliffe, 1993).

Sexuality and space

While the focus of social change is on gendered, racial and ethnic debates within Latin American political geography, there is also the need to stress the significance of sexuality and space. The influence of Catholicism and its homophobia within Latin America has ensured that homosexual and bisexual identities remain 'hidden' within urban and rural spaces. They are marginalised and congregate in effectively 'underground' spaces such as bars and clubs, often in less central locations or in public spaces such as swimming pools and parks, what Edward MacRae (1992) refers to as homosexual 'ghettos'. Spatial references of homosexual identity, of being either 'in(side)' or 'out(side)' of the 'closet', consolidate the question of space and how space is used by homosexual and bisexual groups relative to heterosexual groups. However, although 'ghettoisation' is the norm, there are important diversions. From as early as 1970, when a homosexual organisation was established in a working-class suburb of Buenos Aires, setting in place the future Argentinian homosexual community (1983), Latin American gay organisations have spread throughout the continent, clandestinely where recognition of homosexuality is still anathema to *macho* culture, and more openly in other contexts (Green and Asis, 1993).

One of the most well-known and well-publicised Latin American cultural events is the annual Mardi Gras carnival in Rio de Janeiro. The carnival is an important expression of popular culture, principally of Afro-Brazilians and the *favela* inhabitants of the city, but there are other significant expressions relating to sexuality. For several days, the central city space of the best known city in the largest Catholic country in the world brims with displays of sexuality, with various degrees of nakedness and transvestism, what MacRae (1992) terms the 'carnivalization' of gender roles. The complexities of gendered power relations in terms of carnival performers, costumes and performances are explored by Clare Lewis and Steve Pile (1996, 39) who point to the different interpretations that may be made of carnivals: 'In certain places and at certain times, carnival may be a ritualised resistance, or it may be a contested territory, or it may be a site of hybrid ambivalences, or it may be an opiate of the people.' The diversity of public Latin American carnivals, festivals and fiestas is representative of all of these characterisations.

For Rio de Janeiro's Mardi Gras, carnival may be interpreted as a celebration of Brazilian sexualities in opposition to the construction of sexuality embodied within the Church. It is contradictory and a challenge to the traditional pivots of Latin American society, the Church and the state. However, carnival is a short-lived expression and it is more likely within Brazil and elsewhere in Latin America that public spaces and 'ghettos' will be subjected to 'clean-up' operations as homosexuals are identified and rhetorically merged with sex workers as undesirables. The dominant social ideologies of heterosexuality and patriarchy lead to the spatial marginalisation and persecution of social 'Others'. These groups are seen as a threat to the predominant social fabric and are thus 'hidden' within public spaces.

An important issue incorporated within male homosexuality is the Latin American social structuring of *machismo*. One of the first criticisms of the São Paulo SOMOS group (We Are – Homosexual Affirmation Group) established in 1979 was the criticism of the reproduction of *machismo* in homosexual relationships (MacRae, 1992). This was an important gesture in that it constituted a resistance to the predominant power structuring within Latin American societies. For lesbians this meant resistance to experiences of *machismo* against women in heterosexual spaces and also between partners within homosexual relationships, groups and spaces.

With displays of affection and tactility between and across genders being hallmarks of most contemporary Latin American cultures, homosexual partnerships have experienced a degree of freedom that may not exist elsewhere. However, it is clear that these are carried out in heterosexualised spaces. While heterosexual kissing in public spaces is commonplace, same-sex kissing is considered 'anti-social' and subversive within the constructions of the dominant religion and the patriarchal state. Constructed by homophobics as the 'gay plague', the emergence of HIV/AIDS in Latin America, especially strongly in Brazil, has been a further obstacle in terms of the struggle for homosexual groups seeking to gain access to public spaces on an equal footing with heterosexuals. In Cuba, homophobia feeds directly into the HIV/AIDS medical programmes in that gay HIV/AIDS carriers are spatially confined within enclosed complexes. This reveals both the predominance of the homophobic heterosexual society and also a flawed understanding of the medical condition and its associated medical care.

Spaces are sexualised in the same way that they are gendered, racialised and politicised. Only through recognition of the social complexity of spaces and these considerations in terms of organisation, planning and control can spaces be democratised in terms of power relations and issues of social justice be confronted. Arturo Escobar (1995) writes of the cultural hybridisation of Latin America that encompasses manifold and multiple modernities and traditions. This cultural hybridisation has led to the emergence of multiple spaces of action and interaction within which these cultural identities exist or co-exist. All groups with common identities, along whatever lines, are directly involved in constructions of power relations within society and the transformation of spaces.

RACE, ETHNICITY AND SPACE

The legacy of racism

Bartolomé de las Casas (1474–1566), the Dominican Bishop of Chiapas, offered the earliest written opposition to the methods of conquest and control employed by the *conquistadores*. As a text written as a Spanish criticism of Spanish barbarity, de las Casas reveals the wayward justification of racial superiority employed by the *conquistadores*. This presumed racial superiority became the framework for the acts of cruelty and murder which were embedded within the 'Sacred Rebellion' against the idolatrous and the imposition of the 'natural law' of Christian monotheism (Espinosa Fernández de Córdoba, 1989). De las Casas became in effect the first advocate of Amerindian rights. He described the Spanish conquests as 'evil and essentially anti-Christian' (1995, 108), criticising the need to conquer in order to evangelise, and criticising the construction of the Indian as ignorant, idolatrous, indolent and barbarian relative to the well administered, populous cities that had been 'discovered', located within 'properly organised states, wisely ordered by excellent laws, religion, and custom' (1995, 108). There were only a few other humanitarians like de las Casas, such as Vasco de Quiroga and Antonio Vieira, and Gonzalo Guerrero who fought alongside the Amerindians (Galeano, 1991). The eighteenth-century geographer Murillo Velarde also noted the excesses of the Spanish in their suppression of the Amerindians in his work, comparing them with the Roman and Greek empires. Tzvetan Todorov (1984, 251) reflects that 'The exemplary history of the conquest of America teaches us that Western civilization has conquered, among other reasons, because of its superiority in human communication; but also that this superiority has been asserted at the cost of communication with the world.'

159

Indigenous groups throughout the continent suffered at the hands of the Iberian *conquistadores*. Even in cases of relative success, of the belligerent Mapuches in Chile during the Arauco Wars for instance, or the Indian Wars in the Argentine pampas, the conflicts were debilitating. Despite these conflicts, however, it was not warfare that delivered the genocidal blow to Amerindian societies, but disease. The 'Columbian exchange' brought Asian and European diseases to the Americas, while only herpes was exchanged and introduced into Europe. Typhus, smallpox, measles, influenza and other viral diseases killed hundreds of thousands of indigenous people within fifty years of Columbus's arrival.

In the face of such rapid decline in numbers, it was not until the 1780s that a significant rebellion against the *conquistadores* took place. Tupac Amaru II, a *kuraka* from the Cuzco area of Peru, mobilised a force in the pursuit of an Amerindian/*mestizo* republic which would be able to counter the abuses inherent within the colonial system. Supported by Indians, *mestizos* and black former slaves, the rebellion was overcome by the Spanish forces a year later, leading to a backlash against Amerindian culture (Reid, 1984). Quechua as a language was banned and Amerindian leaders lost much of their autonomy within the colonial system.

The conquest led to movements of racial groups around the continent. For example, indigenous groups from Peru, including black slaves, moved south into Chile. The black slaves were treated better than the Amerindians as a result of General José Miguel Carrera's government legislation that gave free status to the children of slaves born in Chile. Many, however, did not acclimatise and died of tuberculosis and alcoholism. Indigenous group numbers continued to decline rapidly after the end of the seventeenth century (Carvallo Ortiz, 1988).

Racial division was an essential component of the colonial period and this continued into the post-independence period, in terms of economic activity, social segregation and spatial organisation. Luzón *et al.* (1990) use the concept of an ethno-social anatomy of space to explain the organisation of Havana in the late nineteenth century between the white Spanish, white creoles, *mulatos* (mixed race) and blacks. The city was comprised of racial zones reflecting the predominant socio-economic conditions of these groups. However, there were no specifically black areas of the city, since blacks were largely excluded from urban areas except in their occupational roles as domestic slaves or maids.

The struggle for cultural independence in Latin America was closely tied to the political and economic independence struggles. Emancipation, as it was hailed, was to occur in all realms of life. These struggles were epic and focused on the heroic exploits of Bolívar, San Martín and other liberators, providing a common ground of nationhood for the nascent Latin American states. However, the search for post-independence culture after three hundred years of colonialism and a mix of indigenous and European inhabitants and cultures was to lead to the need to construct a national culture based on pre-Columbian and European ideologies. This national culture was subsumed within a broader idea of Americanicity, based on the experience of the Americas (López de Mesa, 1988; Quijano and Wallerstein, 1992).

In all cases, the socio-cultural construction was Iberoamerican rather than Amerindian, marking a new phase of racial colonialism for indigenous groups that is still apparent in the late twentieth century. Emancipation, a Christian concept sailed over from Europe, meant just that for white and *mestizo* Latin Americans, but little for traditional Amerindian groups. The quincentenary protests that brought indigenous groups together from throughout the Americas did not make a distinction between pre- and post-independence. For them, European influence has remained much the same. Poverty rates among indigenous groups continue to exceed those of other groups.[4]

From early independence a well-defined post-Iberian class structure developed. An aristocracy of titled European blood, with traditional aristocratic symbolism such as coats of arms, emerged (Carvallo Ortiz, 1988). The constitutions of the Latin American republics also hark back to independence, giving credence to independence iconography, which has also changed through time, such as the national hymn, the national coat of arms, and the flag (de Dios Carmona, 1987). The continuity of the nobility was ensured by intermarriage and dominance of institutional power via the control of positions within the army and the Church. In Chile, landed wealth was kept as the preserve of these élites via *mayorazgos*, whereby estates and property could not be sold and were passed to the eldest son. The outcome was that a *nouveau* nobility became established in post-independence Latin America, maintaining a certain 'racial hermeticism' (Carvallo Ortiz, 1988, 68), and confronting attempts to establish truly democratic republics. The independence leaders such as Bolívar, San Martín, Artigas and O'Higgins came into conflict with these new élites, and lost.[5]

Black and Amerindian Latin America

Within this élitist social hierarchy, racial and ethnic 'Others' became marginalised. In Latin America the social constructions around the notions of blackness, Indian, and *mestizaje* (mixed race) have been fundamentally important in terms of the structuring of Latin American social hierarchies, despite the fact that it is almost solely the latter that has been utilised for the construction of a mixed-race Latin American nationalism.

Colombia and Brazil are the countries with the largest black populations and these are well-established on a regional basis, in Brazil's North-east, and on Colombia's Pacific coast. The Chocó province of Colombia has an 80–90 per cent black population (Wade, 1993). Peter Wade uses the idea of a cultural topography with which to interpret Colombian national space with its distinctive associations: black – coastal; white/*mestizo* – interior; Amerindians – lowlands. He also goes on to note that it is the dominant groups that dictate these national cultural geographies and have the power to overcome them or to ensure their persistence.

The construction of Colombian nationhood incorporates ideas of territorial, moral and racial integration linked with notions of the city and the frontier (Wade, 1993). Colombian blacks have consistently suffered from a refusal to being granted ethnic status due to these predominant constructions of the nation-state. Between Amerindians and blacks there is also antagonism, since Amerindian groups are perceived, and perceive themselves, as being local and traditional, failing to acknowledge the long history of black culture within the region. Even within the new 1991 constitution, the situation of blacks within the pluri-ethnic, multicultural character of Colombia (as defined within the constitution) remained unestablished (Arocha Rodríguez, 1992).

In Brazil, Afro-Brazilian movements now number over 600, emphasising education, black popular culture (such as the Samba tradition), and Afro-Brazilian religions such as Candomblé and Umbanda.[6] The most significant umbrella organisation is the São Paulo-based Unified Negro Movement (MNU), which has promoted the black history of Brazil, highlighting rather than ignoring the impact of slavery in the creation of contemporary Brazil. The greatest success for the Afro-Brazilian popular movements was the 1988 constitutional amendment, which made racial discrimination illegal (Burdick, 1992). Despite the legislative developments in favour of Afro-Brazilian recognition, the roles of Samba and carnival have succeeded over a longer time-scale in wedding Afro-Brazilian popular culture to Brazilian national culture, raising black consciousness within the national population (Rowe and Schelling, 1991).

As with Latin America's black populations, Amerindians have experienced similar power-space constructions that have been put in place in order to marginalise them. Amerindian groups often transcend modern nation-state frontiers and, as transboundary social groups, are considered to complicate the nation-state's drive for territorial and social integrity and unity.[7] Examples of such racial groups are the western Amazonian tribal groups, the Guaraní of Paraguay and Argentina, the Guajiro of Colombia and Venezuela, the Aymara of Peru, Chile and Bolivia, and the Maya of Mexico, Guatemala and El Salvador. Nation-state constructions have led to their continued persecution in the post-independence era. The genocide of Amerindian groups is not a historical fact but an ongoing one. In 1979, the UN Human Rights Commission, with reference to the Yanomami, announced that the largest Amazonian Indian tribes would cease to exist in twenty years. The Nambiquara, 15,000 at the turn of the century and 2,000 in 1990, provide a benchmark (Galeano, 1991).

While indigenous groups are most often marginalised within the nation-state and its policies, the same holds true for many late twentieth-century development projects financed by multilateral institutions. In nine special development projects established in Peru in the 1970s and early 1980s by the private sector, the World Bank, the Inter-American Development Bank and USAID, the fifty-six ethnic groups of 220,000 Amerindians within the development areas were largely ignored. Peru's President Velasco respected Amerindian rights and needs, prohibiting land grants to commercial enterprises, but subsequent presidents, Morales Bermudez and Belaúnde, submitted to stronger sectoral interest groups to open up Amerindian territories to commercial development (Reid, 1984). The Peruvian story is symptomatic of the longer history of failure to recognise indigenous land rights and human rights. The only avenue left open to these groups has been resistance.

Indigenous resistance

Ecuadorean indigenous groups have consistently opposed the nation-state. The first national Indian Federation was established in 1944 and persistent land reform pressures, alongside peasant groups and Alliance for Progress objectives, led to the Agrarian Reform Laws of 1964 and 1973. During this time other peasant organisations were founded, creating a large interlinked indigenous peasant network of groups, bringing together the peasants of the coastal regions and the indigenous groups of the eastern regions. With objectives such as bringing agrarian reform under peasant control, abolishing the restrictive 1979 Agricultural Development Law, expelling religious and 'development' organisations from Amerindian areas, ending repression in rural areas, integrating peasant organisations, and demanding respect for Amerindian culture, the peasant-indigenous movements have continued to exert pressure on the government by mobilising approximately 60 per cent of the rural population (Corkill and Cubitt, 1988).

In May 1990, the Confederation of Indigenous Nationalities of Ecuador (CONAIE) initiated a land rights insurrection that mobilised a large percentage of Ecuador's 40 per cent Amerindian population. By June the highland provinces were paralysed and the government's Agricultural Reform Ministry (IERAC) was forced into negotiations relating to land distribution and communal land rights. The insurrection became known as the sixth Indian uprising, following similar major confrontations in 1578, 1599, 1615, 1766 and 1892 (although CONAIE acknowledges over 145 insurrections since the Spanish conquest) (Field, 1991). With other organisations besides CONAIE involved in indigenous issues, such as the Federation of Ecuadorean Indians (FEI), the Federation of Peasant Organisations (FENOC) and the Confederation of Indigenous Amazonians (CONFENIAE), indigenous opposition to the dominant state apparatus is still strong, as exemplified by the 1990 insurrection. With other similar peasant and Amerindian insurrections in other regions of Latin America, support for indigenous claims increases and international pressure is also brought to bear on Latin American nation-states.

Many Amerindian groups throughout the continent have suffered from exposure to European-based socio-economic and cultural systems. The Ecuadorean Shuar Federation was established in 1964 to confront the increasing commercial interventionism in Ecuador's eastern regions. The Ecuadorean government's objectives in the region were more to do with territorial integrity to counter Peruvian claims in the region than with seeking an understanding of existing Amerindian spatial organisation. The Ecuadorean Institute for Agrarian Reform and Colonisation (IERAC), established under the US-backed Alliance for Progress (1964) and the Institute of Colonisation of the Ecuadorean Amazon Region (INCRAE) (1978), provided the structures for the commercialisation of the region and its settlement (Villarreal, 1986; Corkill and Cubitt, 1988). The lack of success in achieving agrarian reform has led to a greater attention to agrarian development. While this may increase productivity, landholding structures remain relatively unchanged.

Geopolitical objectives and commercial exploitation have characterised interventions into indigenous regions and the imposition of state-centric territorial-administrative structures. These structures have been super-imposed on traditional social structures and organisation. In the case of the Shuar, centres of 25–30 families each organise local spaces for production (Corkill and Cubitt, 1988). It is through the activities of the Federation that Shuar culture, identity and land rights have been reasserted to counter this internal colonialism.

The state has to consider how indigenous groups are inserted into, or recognised independently from, the structures of the nation-state, with respect for their existing organisation and rights rather than in opposition to them. Ultimately it is only through recognition of linguistic and cultural diversity, an understanding of the ecological management of these groups, and a reinterpretation of geopolitical and resource frontier concepts that republican nation-state and indigenous perspectives on development can be accommodated rather than fought over. This process demands a change in cultural attitude towards indigenous groups. Eduardo Galeano (1991, 17) highlights misleading contemporary cultural perceptions:

> Throughout America, from north to south, the dominant culture acknowledges Indians as objects of study, but denies them as subjects of history: The Indians have folklore, not culture; they practice superstitions, not religions; they speak dialects, not languages; they make handicrafts, not art.

The perspective of many indigenous groups relative to contemporary nation-states is encapsulated in the 1990 declaration of Guatemala's Committee for

Campesino Unity with regards to the Quincentenary of Columbus's arrival in the Americas (see Box 5.1).

Latin American political spaces are increasingly multicultural as one moves along the rural to urban spectrum. One may say that to find indigenous Amerindian groups one has to seek out the rural communities in which they are still engaged in traditional activities, since in the largest urban centres Amerindians co-exist in terms of society, economy and culture with *mestizo* and immigrant groups. Urban centres act as the cultural melting-pots within which local identities become cosmopolitanised. While the Central American and Andean urban centres still reflect the largely Amerindian composition of the national populations, within the Southern Cone and Brazil racial and ethnic boundaries are transgressed as people engage in urban socio-economic activities and submit to the predominant cultural images projected by television, radio and retail marketing (Rowe and Schelling, 1991).

Box 5.1 Declaration of the Committee for Campesino Unity, Guatemala

'On the 498th anniversary of the Spanish invasion of the Americas and our Mayan lands, we, indigenous and ladino campesinos, organised in the Committee for Campesino Unity (CUC), address all Mayan peoples, indigenous groups, trade unions, popular organisations, all Guatemalans, indigenous people throughout the world and the international community. Declaring:

1   498 years have passed since that fateful 12 October 1492 ...
2   On the 498th anniversary of the invasion, we must also celebrate this widespread and courageous resistance, thanks to which our roots remain unbroken and our identity and culture has not been destroyed ...
3   We look to our brothers and sisters in the Communities in Resistance in northern Quiché, as a living example of the difficult history and condition of today's indigenous and poor ladino ... This 12 of October 1990, 498 years after the invasion, we call on all ethnic groups to make the day one of reflection, analysis, struggle and resistance so that united and organised, our people can confront the hunger and misery and fight for respect and dignity, our culture and identity and for the recuperation of our stolen lands ... We call on ethnic groups and popular organisations to turn the campaign into a celebration of 500 years of indigenous and popular resistance. We hope that this campaign will be a space for reflection in the American continent, a project for unity, an analysis drawing together past and present to create the foundations of a project for the future on which we can build day by day with strength and decision.'

(Guatemala, 11 October 1990)

Source: Koning (1991)

Resisting the 'whitening' images of projected mass culture, blacks and Amerindians have to reflect back to Bartolomé de las Casas (1995, 110) and his defence of their culture at the initiation of this process: 'they are docile and clever, and in their diligence and gifts of nature, they excel most people of the known worlds.' Peaceful, intelligent, diligent ... clearly the hispanic construction of civilisation could tolerate none of these attributes without conversion to Christianity or slavery. The resistance to this 'civilising mission', now 500 years old, has been bloody and painful but will surely last until the European and US exploitation has been completed, Latin American societies have been 'whitened', and the mapping of cultural spaces has become less diverse and increasingly uniform. Rigoberta Menchú, as a representative of indigenous resistance throughout Latin America, but particularly Guatemala, has struggled against this process and raised the profile of this struggle throughout the world to great effect.

Political geography must acknowledge that spaces have multiple characteristics, culturally and socially as much as economically and politically. Political geographers must avoid falling into the trap of extinguishing diversity in order to simplify analyses of power and space.[8] Only a recognition and tolerance of 'Others' throughout Latin America's territorial spaces can lead to a peaceful process of long-term socio-economic development.

Box 5.2 Rigoberta Menchú, Guatemala and indigenous opposition

In 1992, Rigoberta Menchú Tum, a Maya Quiché, won the Nobel Peace Prize for her work with land rights movements, the rights of indigenous Maya, and the rights of the urban population for higher standards of living. The history of her struggle is a history of opposition to the authoritarian tradition of Guatemalan politics and the role of the USA in preserving that tradition. From 1954 when the CIA backed a coup against the land reform administration of Arbenz, it has maintained its high level of influence within the country. An example was the withdrawal of support for the violent General Romeo Lucas García (1978–82) which led to the administrations of the more favoured allies, General Efraín Rios Montt and General Oscar Mejía Víctores. These military administrations have engaged in consistent campaigns against opposition forces and within these campaigns the systematic persecution of indigenous peoples. By countering Mayan culture as a counter-insurgency measure, the military has attempted to reorganise the highland areas, forcing population movements to centres of control, legitimised as poles of development, and controlling all rural 'development' projects. The establishment of Civic Patrols (PACs) in 1982 also led to violent confrontations as social groups confronted one another, spurred on by the military.

As a response to these authoritarian regimes and their policies, social movements were established to press for land reform and human rights. The Mutual Support Group (GAM) and Coordinating Council of Guatemalan Widows (CONAVIGUA) established human rights opposition, while the Pro-Tierra Movement initiated land occupations and provoked the government into land partition and distribution. The Quincentennial 'celebrations' in 1992 also provoked indigenous opposition into organising on several issues. Rigoberta Menchú was the figurehead of this opposition.

As an active member of the Peasant Unity Committee since 1979, Rigoberta Menchú was forced into exile in 1981 after participation in the activities of the January 31 Popular Front following the regime-related deaths of her parents and brother. In exile, Elisabeth Burgos helped write her oral history, *I, Rigoberta Menchú*, which was projected onto the international stage as a representation of the struggles of Guatemalan indigenous groups. The book and the ensuing publicity of the Nobel Prize brought considerable pressure to bear on the Guatemalan regime, which had arrested her in 1988 on her return to the country. She was released only after a government amnesty for guerillas.

Rigoberta Menchú is a leading figure in Latin American political geography. By bringing to international attention the transformation of highland Maya economic and cultural activities and the role of the military in establishing new territorial policies to combat opposition forces and Mayans themselves, she has emphasised the changing power-space relations of indigenous groups throughout the Americas but most particularly those in Guatemala during the 'lost decade' of the 1980s. Women have become increasingly vociferous in terms of the pursuit of issues of social justice throughout Latin America since the 1970s. Rigoberta Menchú is the tip of the iceberg of Latin American women's political activity.[9]

Sources: Handy (1992), Brittin (1995)

THE CONSTRUCTION OF NATIONALISM

Creating nations

The construction of the republican nation-states at the beginning of the nineteenth century also entailed the development of a national consciousness and a national identity. With a range of black, Amerindian, *mestizo* and creole citizens within the proclaimed republics, a long process of social construction was put in place in order to establish commonalities across existing classes and races, structures founded upon European and American origins.

In many ways, the process can be seen to have been unsuccessful. What must be made clear is that independence was not an uprising of popular sentiment culminating in revolution. Latin American independence was steered by the Latin American élites, essentially *criollos* (creoles), those who had been born in Latin America but originated from European families. These *criollos* sought independence from the *peninsulares*, those still closely associated with the Iberian peninsula. These associations manifested themselves in terms of commerce, administration and family ties.

It is for these reasons that late twentieth-century Latin American citizens are likely to be noted more by racial and ethnic origins than by their national identity. There are more commonalities among Amerindian

Plate 5.1 The statue of San Martín, Argentina's 'liberator', occupies a prominent
position in front of the presidential palace in Buenos Aires' Plaza de Mayo.

groups wherever located within the continent, and between citizens of
European descent, than there is sense of national identity between them.
The most obvious evidence of this situation is the spatial organisation of
these groups within national space and their distinct levels of segregation,
at regional and urban levels.

Plate 5.2 This monument, also in the Plaza de Mayo, commemorates the date of independence.

Despite this most important caveat, the Latin American republics can be said to have established their own national identities based on post-independence iconography and traditions. The role of independence within these identities is paramount. Independence provides the starting point of

169

republican nationalism. Heroes of the independence movement are considered to be the founders of the nation, independence battles and the actual independence day are the key dates around which the historical construction of nationalism is based. Geographical landmarks such as monuments and pivotal roads and avenues also proclaim this independence legacy.

Clearly the significance of the independence struggle as the starting point of republican history has had a strong negative effect on the recognition of pre-Columbian histories and the role of Amerindian groups within the national territories. While independence certainly changed the ways in which Amerindians were acknowledged within the continent, the model of the nation-state, imported from Europe and merged with the specifics of creole élite cultures within the continent, did little to alter the process of Indian acculturation and exploitation. For Amerindians, independence did not signify a return to pre-Columbian times, but rather a new flavour of 'Americanised' European élite domination.

A hundred and fifty years on, Amerindian groups are still struggling for greater recognition of their identities within the national identity that does little to reflect their own cultural histories. They have been subsumed into a European-based identity that they are largely unable to identify with. Geographical and historical education structured at a national level attempts to suppress indigenous cultural histories in favour of post-independence republican histories (Escolar et al., 1993; Radcliffe, 1996a, 1996b; Radcliffe and Westwood, 1996). The issue of recognition of the diversity of Amerindian cultures is seen as a de-unifying problem that has to be overcome with a nationwide version of history that makes the links between all those living within the national space.

Constructions of national history 'hide' regional, racial and ethnic differences in the pursuit of objectives of national integration and common purpose. Sarah Radcliffe (1996a, 37) makes the point that 'no hegemonic national identity has been built around a shared imaginative geography' and that 'popular geographies and identities continue to rearticulate identity, place, and nation'. Another important point is that by starting the republics and their construction at independence, the 1810 *uti possidetis* boundaries remain embedded in national consciousness. These boundaries run roughshod over delineated indigenous spaces, which are left unrecognised in the same way that the *conquistadores* and those that followed during the colonial period failed to recognise them.

The Amazon basin provides a good example of the diversity of groups subsumed within national histories based on independence and post-independence. According to the Treaty of Amazonian Cooperation, there are 900,000 people from 300 linguistic groups in the Amazon basin. In Ecuador, for example, the eastern province is inhabited by numerous minority ethnic groups, such as the Aucas, Cofanes, Sionas, Secoyas, Tetetes, Quichuas, Asuaras, Zaparos and Cayapas (Navarro Dávila, 1985). The same diversity is seen in Peru, Brazil and Bolivia. As Amerindian groups, their associations are more likely to be with other Amerindian groups in terms of culture, production and reproduction. However, they are citizens of nation-states with histories which have no resonance with their experiences and beliefs.

What is required in contemporary Latin American nation-states is to establish recognition of Amerindian groups as an integral part of the American nature of Latin American nationalism. A recognition of diversity, rather than the pursuit of uniformity most strongly sustained under authoritarian regimes, is the only way that a mobilisation of national conscience can lead to recognition of and rights (particularly human rights) for these groups, rather than in their integration and absorption into an alien structuring of national space based on traditional European élite conceptualisations of spatial organisation.

The particular nature of the emergence of Latin American states, from a fracturing of external relations within a short time, makes it difficult to apply the Marxist theories of nationalism associated with uneven development and the nature of production (Nairn, 1977; Soler, 1980) nor Anderson's important prerequisite of print capitalism (the increased circulation of printed literature) in order to act as the mechanism of diffusion of nationalist fervour (Anderson, 1991). The republics clearly were keen to establish independence, severing the Iberian ties that had drained the continent of its resources for three hundred years. However, issues of identity and self-determination were to play key roles.

The promotion of nationhood

Unlike many independence struggles, in those of Latin America language was not a factor on which conflict was based. Despite the predominance of Amerindian languages throughout the continent, the independence movement was not a racial or ethnic struggle, rather it was the outcome of a breakdown of an overextended empire suffering pressures close to home and unable to counter the divisions from within. In many ways, there are overtones of the fracture in the Inca empire which facilitated the progress of the Iberian *conquistadores*. Using the same analogy, the *conquistadores* utilised the support of discontented *kuraka* as the *criollos* used Amerindian support centuries later.[10] The advantage of incorporating Amerindians and blacks into the independence movements was that it reduced their opposition to the movements.[11] Nevertheless, rhetoric relating to the recognition of non-white and non-*mestizo* cultural groups went only a small way towards improving the rights for these groups and their benefits from productive activities. These groups were attached on to a creole nationalism, rather than integrated into a broad-based Latin American nationalism.

For Amerindian groups in the Andean region, their independence struggle had taken place the previous century with the uprising of Tupac Amaru against the Spanish. Although unsuccessful, Tupac Amaru had provided a model of insurrection for indigenous groups. Toussaint L'Ouverture had done the same as Tupac Amaru for blacks, leading slaves against landowners in Saint Domingue (Haiti, the western area of the Dominican Republic, formerly Santo Domingo) in 1791. Unlike Tupac Amaru, L'Ouverture was successful, which led to Haiti's early independence from France in 1804 and to a similar insurrection in neighbouring northern Venezuela (Robinson, 1990). While independence meant independence from Spain, and Portugal in the Brazilian case in 1889, racial dependence and power relations derived from the racial structuring of colonial societies were essentially retained.

In terms of the timing of the independence movements of the early nineteenth century, several factors came to a head simultaneously. The 'air' of revolution sweeping through European and US political ideologies provided the catalyst. The North American Revolution in the 1770s was followed by the 1789 French Revolution and the wave of republicanism that was created provided a focus for opposition to the tax regimes and other impositions of the Spanish king, Charles III (1759–1788).[12] The creole élites were established in their own right as Latin Americans by birth, which had reduced their affiliation to Spain the empire, as controlled from Madrid. They took umbrage at the increasingly demanding monarchist interventions, and to their social construction as inferior Spaniards by the *peninsulares*. Republicanism provided the ideology to justify a breakaway.

This Spanish American republicanism adopted many of the territorial-administrative features of republican France, such as territorial sub-divisions and the appointment of prefects who were responsible directly to the central government or the president (Robinson, 1990). These French republican features were employed alongside particularly Spanish American structures that accommodated existing power relations between *criollos* and *peninsulares* and Amerindians. In terms of French political-organisational influences and British commercial influences in the immediate post-independence period, a new wave of post-Iberian European relations was rapidly put in place.

As a process taking place alongside the construction of 'modern' nation-states in Europe, the question is how a sense of national identity could be established where the predominant social formations had been established by associations through a hierarchy from the *encomienda* system to the vice-royalties to Madrid. It was precisely the colonial territorial-administrative divisions that lent coherence to the movements, and also prevented Bolívar's dream of a unified continental system. The audiences and vice-royalties had established territorial hubs of activity within Central and South America. Societies were controlled by a strong centralised system emanating from hubs such as Lima and Buenos Aires. It was from these hubs that the construction of nation-states and nationalism issued forth from the republicans, in the same way that subservience to the Spanish crown had emanated previously.

In the case of Argentina, the creation of a national community involved the development of the following (Dodds 1993c, 316): the securing of a form of political-economic authority among the provinces; the establishment of a set of *international* relations; and the construction of an 'imagined community' of creoles in a demarcated territory. The indigenous groups of this region were considered more of a threat to the establishment of new territorial-administrative apparatus than as a part of it. It was not until the latter half of the nineteenth century that Argentinian republican nationalism was extended throughout contemporary Argentinian national space (Escudé, 1988). Geography was very important in this process as territory was established as a patriotic symbol, hence 'patriotic territory'. The constructed notion of patriotic territory, disseminated via geographical education, served to promote a national geopolitical consciousness. This conjoining of nationalism and territory, what Escolar *et al.* (1993, 362) call 'territorial mythology', has been significant throughout Latin America since it became a justification for the negotiation of or claims over boundary disputes.

In the same way as Argentinian nationalism passed through phases of development of a national consciousness and the consolidation of national territoriality, other Latin American republics experienced similar processes. For instance, Venezuelan nationalism passed through three broad phases: the initial phase prior to independence, when nationalism became defined as a creole aspiration; a second phase from 1850 to 1900 that gave rise to the development of Venezuelan political and cultural nationalism; and the third phase of economic nationalism and modernisation in the early twentieth century based on oil wealth and the urbanisation of the national population (Dávila, 1993).

During the post-independence period and into the twentieth century, the construction of Latin American nationalism has been consolidated, with differing timetables and internal conflicts according to the particular republics. The struggles among regional *caudillos* slowed the process considerably during the nineteenth century, but by the turn of this century European immigration and the urbanisation process acted as a centralising force, weakening regionalism to a large extent. Increased communications and changes in production also assisted in this territorial and ideological integration process.

In the late twentieth century, Latin American republics have well-defined, but certainly not inclusive or democratic, constructions of national identity, iconography and national pasts. International disputes and conflicts serve to settle or renegotiate real and imagined boundaries that are part and parcel of these constructions. As with the creole impulse for independence, contemporary nationalism falls into the realm of a continuation of a European-based project for territorial organisation and control by landowning, commercial and now industrial élites, to the exclusion of the diversity of cultures existing within territorial spaces bounded by the nation-state.

Apart from racialised and class-based considerations, there also needs to be a recognition of the gendered constructions of nationhood and the nation-state. Sarah Radcliffe (1996a, 6) advances the idea that for a broader interpretation of the nation-state, beyond the purely political, nations should be understood as 'multiethnic, diasporic communities constituted through relations of difference and globalisation (economic and cultural), and through the practices of nation-building and imagination'. In this way, nationhood may be interpreted in its widest sense, not solely in terms of a centralised construction of national identity, through educational curricula for example, but also in terms of social identities (gendered, racialised, sexualised) and wide variations in discourses and practices surrounding nationhood and national identity.

MOVEMENTS, COMMUNITIES AND POWER

The mobilisation of civil society

In the same way that indigenous groups have mobilised against the state for human and land rights, other social groups have mobilised for various political, economic and social goals, from campaigns against human rights abuses to basic needs projects such as food kitchens. Social movements have always existed within Latin American societies. They have served to stimulate change, leading to varying forms of social and economic development.

While previous social movements should be acknowledged, the impact of the New Social Movements (NSMs) in Latin America based on resistance in the 1960s and continuing more strongly into the 1980s has been profound. They have redefined power relations which, in turn, have been reflected spatially. They have been considered new in that they were novel within the Latin American context: new forms of struggle against repression and subordination in capitalist society; new in terms of the autonomy of the subjects rather than in terms of economic class struggle; and new in terms of what David Slater (1985, 7) terms 'basis-democracy', whereby people became more involved in group decision-making and cooperation.

The common basis of the diverse types of movements (from human rights to sexuality to ecology) has been their stressing of popular participation and its empowering nature, also their opposition to and pressurising of the state. They are characterised by their internal democracy, grassroots participation, open meetings, collectivisation, the rotation of leadership, and administrative transparency, although these criteria vary among movements. What the social movements have offered society is a different civic political culture (Fals Borda, 1992), divorced from or at least at odds with traditional institutional politics.

The rise of social movements on a widespread basis from the 1960s has

Plate 5.3 This publicity promotes the 'Week of Chilean-ness', encouraging greater participation in events organised around the date of independence. Santiago (Chile).

Plate 5.4 This vast topographical representation is displayed in a central park in Guatemala City, to promote awareness of the Guatemalan territorial area and the location of places.

played an important role in Latin American political geographies in that the state has been challenged and more people are engaged in popular mobilisations and activities, leading to significant changes in the organisation of local spaces, in representation and rights as citizens of the state (Roberts 1995, 186):

> Social movements in Latin America are the most visible signs of the struggle to define and redefine citizenship. Their field of action differs from what it is in the advanced industrial countries because in Latin America the project of modernity is far from accomplished, particularly in terms of formal democracy and an adequate standard of living for the majority of the population.

Social movements are diverse in terms of objectives and size. The reasons behind mobilisation are also diverse. They may be related to employment, markets and the economy, to racial and ethnic conflicts, to gender and sexuality based issues, to political defiance, to human rights, or to religious movements, among many others (Eckstein, 1989). They are also significant in that they transform the political to the personal. The political becomes localised, direct and less institutionalised. In many ways, the human rights campaigns against the abuses of the Latin American dictatorships of the 1970s and 1980s revealed the personalisation of the political (Navarro, 1989). People have become empowered and politicised at the local level, confronting the failures of institutional politics to overcome many social, cultural and economic obstacles that face numerous social groups within Latin American society.

Within communities and at the grassroots of civil society, it is women who often initiate the greatest change and action. This is very much in line with the idea of the 'triple day' that faces most Latin American women: they are economic producers within the public sphere; they are social reproducers within the private sphere of the house; and they also maintain communities. It is the local, neighbourhood informal networks established most often by women that constitute the links, ties and associations of support,

Table 5.1 Ideological, organisational and institutional development for new social movements, 1970s–1980s

1	Progressive Church liberation discourse and pastoral activism.
2	Self-criticism of the left and re-evaluation of strategies of social change from the 1960s.
3	Generation of new international networks among urban and rural residents from different social sectors.
4	Massive expansion of the developmentalist state itself and the role of state agents in encouraging their clients to demand social services.
5	Targeting new groups, particularly women and indigenous peoples, by national and international development establishments.

Source: Alvarez and Escobar (1992)

exchange and care that characterise community life.

One example of such a community based around women's activities is to be found in Indio Guayas district of Guayaquil, Ecuador, where women have mobilised in a *barrio* committee to establish a strategy of 'godparenthood' (*comadrazgo*) on the birth of their children. Such an important social step towards establishing neighbourhood solidarity has led to increased political activity and popular participation to counter the poor infrastructural characteristics of the *barrio* (Moser, 1993). Under the populist politics of exchanging votes for infrastructure and service provision, Indio Guayas and other Guayaquil *barrios* received waste dumps, electricity and water supply. Under this type of political organisation within local spaces, neighbourhoods are able to mobilise in order to achieve their basic needs provisions. The manifestation of populist politics reveals the importance of neighbourhood 'bloc voting' in the political process.

Beyond groups that focus on the empowerment and activities of women, there are wider movements that are representative of regional identities, bringing together diverse social groups within a broader identity linked with a region. An example of such a movement is provided by the local organisation for political pressure known as the Worker-Peasant-Student Coalition of the Isthmus (COCEI) based in Juchitán, Mexico. As a powerful regional body within the Mexican political system, COCEI provides strong opposition to the PRI. COCEI provides a significant example of a social movement in terms of what Fals Borda describes as an 'ecological-cultural regionalization that reflects that actual dynamics of community life' (Fals Borda 1992, 311).

The Coalition was founded in 1974 to reflect the regional indigenous identity of Juchitán and to respond to the usurpation of peasant lands. From 1974, the movement was involved in active opposition to the PRI, seizing town halls, initiating hunger strikes, and challenging government repression. The outcome of this activity was the winning of two mayoral seats and a large regional support base (Gardy, 1992). As a predecessor to the Zapatistas of Chiapas, COCEI reveals the impact of regionalism within Latin America, based as it is on the legacy of *caudillismo*, economic regionalism and strong regional identities themselves often based on ethnicity.

While NSMs were vocal and active during times of authoritarianism, their new roles within the democratic transitions of the Southern Cone states and Brazil are not clear. The political arena has changed significantly since the mid-1980s and the dualism of state-civil society has been blurred. Some movements still operate outside or 'beyond' the state, others have formed partnerships and collaborate with transitional democratic state institutions and organisations (Reilly, 1995). Entering this new space between state-civil society leads to problematic issues relating to control and power linked to finance and the management of projects.

Plate 5.5 Many social movements have concentrated on the improvement of sanitation and water supply infrastructure. For the stilted houses of Castro on the island of Chiloé (Chile), the water beneath receives the household effluent, at high and low tide.

Contemporary strategies

Clearly, as democratic forms of state control mature and new partnership styles are accommodated, there will be more opportunities for social movements to seek change through the state, building on their own self-help mechanisms. A major obstacle, however, will be the widespread lack of confidence in the state and institutional politics and in politicians themselves (Cavarozzi and Palermo, 1995). This suggests that this process may be a long one. For some movements, it is unlikely that trust in the political process and in the state apparatus will be forthcoming; in these contexts, the 'alternative' strategies of the grassroots will continue beyond the state.

A significant issue in this context is the role of non-profit-making non-governmental organisations (NGOs) in these new processes. Funded principally by benefactors beyond the continent, NGOs have been closely tied to NSMs, popular mobilisations and grassroots projects since their origins in the 1950s, emerging from popular religious responses to social unrest (Macdonald, 1995).[13] The degree to which NGOs sustain local development or simply replace state assistance for basic needs is debated. However, during the crises of the 1980s, NGOs performed a vital support role for Latin America's most vulnerable groups. The question for the 1990s

Plate 5.6 Household effluent flows into a stagnating channel and poses a health risk. Flores (Guatemala).

is how NGOs link with the new democratic state administrations (in partnerships) or remain independent from them, working as previously and also bringing pressure to bear on the distribution of the new social investment funds that have been put in place for local development projects in many countries.

The link between the multiplicity of Latin American identities and local group action is an important one. It is the struggle embodied within these projects that gives rise to changes in space and power relations between these

179

Plate 5.7 While this drainage canal is flushed with water only at times of high rainfall, it acts as a conduit for household effluent, removing waste from the vicinity. La Libertad (El Salvador).

groups and other groups, such as those representative of the state, that is, local government. Political geography uncovers the spatial dimensions. Effectively it maps these resistances and forms of social change that are so important in terms of the politics of the state and its relationship with civil society. Contemporary Latin American mapping must take into account the fragmented social and political spaces that have emerged from the authoritarian and institutional political histories of the republics since independence.

Plate 5.8 Improvements in housing materials and construction have also been targeted by social movements. La Libertad (El Salvador).

The failure of the state to act as a provider and as a non-partisan player within the political system, often usurped by authoritarian regimes to the left and right, has led to dissatisfaction in Latin American civil society. There is a sense of struggle for liberation that has emerged, liberation from the imbalances of capital and from the prevalence of violence (Comacho, 1993). Issues such as human rights abuses, absolute poverty, poor infrastructure, environmental degradation and land rights have all led to local forms of pressure organised within movements. Most often these movements and organisations are locally organised and mobilise around a local issue. However, movements may also exist at a national level and link with similar movements within the global arena. This has been especially clear in the case of human rights and environmental organisations.

In many ways, social movements since the 1970s have been drawn up against the cultural revolution embodied in neoliberalism which almost claims the development high ground as the ultimate truth (Escobar, 1995). Participants in Latin American popular movements might well agree with Zamora (1995, 7) who parodies Marx to declare that 'neoliberalism is the opiate of the business class'. An important issue underlying this type of criticism is the extent to which the market is linked with forms of democracy. An early example is the case of Chile (1970–73), when the forces of neoliberalism (in this case ITT with US government backing in the shape of the CIA) toppled the democratically elected socialist administration of Salvador Allende.

The bankruptcy of traditional political institutionalism has led to a power vacuum that social movements have moved into, offering political alternatives of 'people power' in order for the poorest and most vulnerable sectors of society to achieve their basic needs. Based on pluralism and tolerance, movements emerging from within civil society have had marked effects on Latin American political geographies, moving through and across spaces via socio-geographic networks and coordination mechanisms, while also consolidating spaces under their influence. These movements and organisations are likely to continue to be important in terms of providing efficient, decentralised forms of administrative organisation into the next century, in spite of current transitions to democracy.

FURTHER READING

Tessa Cubitt's *Latin American Society* (1995) is a general introduction to social actors and social change. For greater depth of analysis of cultural and identity issues, see Rowe and Schelling's *Memory and Modernity: Popular Culture in Latin America* (1991), Daniel Levine's (ed.) *Constructing Culture and Power in Latin America* (1993) and Radcliffe and Westwood's *Re-making the Nation: Identity and Politics in Latin America* (1996). The organisation and activities of social movements are detailed in David Slater's (ed.) *New Social Movements and the State in Latin America* (1985), Susan Eckstein's (ed.) *Power and Popular Protest: Latin American Social Movements* (1989), Alvarez and Escobar's (eds) *The Making of Social Movements in Latin America: Identity, Strategy and Democracy* (1992), and Radcliffe and Westwood's (eds) *Viva: Women and Popular Protest in Latin America* (1993).

6

TOWARDS A DEMOCRATIC
POLITICAL GEOGRAPHY
OF LATIN AMERICA

> Throughout its history, geography has frequently cast itself as
> aide-de-camp to militarism and imperialism.
>
> (Livingstone 1992, 352)

DEMOCRATIC POLITICAL GEOGRAPHY

This text seeks to shift a balance to a small degree, by viewing the world from Latin America rather than from elsewhere. However, a more significant shift is required within political geography to ensure that Latin America is not inserted into political geography analyses as a marginal and peripheral player in a larger capitalist game of geopolitical economy. For Latin Americans, they themselves are central to their perceptions of political and geographical worlds. Their personal political geographies around the home and neighbourhood are not peripheral to Anglo-European constructions of political geography, merely different from them, even in some cases very similar to them. It is by engaging with the differences between political geographies on a global and local basis that a better conceptualisation of power-space relations may be established.

In the conclusions of his *Political Geography and the World Map*, Yann Goblet (1955, 291) noted the apparent trend of 'diseuropeanisation' and 'decline of Europe' in international affairs. Forty years later, Europe remains a powerful region in international affairs but its role has changed. Despite this change, a considerable proportion of political geography literature still emanates from the region, creating an imbalance in terms of international coverage based on European perspectives on the world. When material deriving from the USA is added to the European literature, this proportion increases considerably. The outcome is that political geography has become dominated by Northern views of the world, especially in terms of the treatment of nation-states in the southern hemisphere. A narrow focus on the political geographies of Northern nation-states adds little to an understanding of the structures and processes of late twentieth-century global political economy, especially in terms of processes in the South. A similar criticism may be directed at literature on globalisation.

While there is widespread recognition of the decline in ideological conflict in the East–West axis, relatively little ink has been spilt by political geographers on developments in the North–South axis. In terms of political geography's concentration on the East–West rather than the North–South divide, one has only to review the collection of observations by key European and North American political geographers at a conference round table on 'The Political Geography of the Post-Cold War World' (Association of American Geographers, Miami, 1991). The South is notable by its low profile in their analyses.

The traditional point of contact of political geography with the southern hemisphere has been in the realm of imperialism and colonialism, or in the relationships of cores and peripheries (Quijano and Wallerstein, 1992; Taylor, 1993b). The use of this simple geographical relationship would appear to be as common in post-Cold War analysis as it was previously in investigations of dependent capitalist relations between North and South (see Frank, 1969; Cardoso and Faletto, 1972). The emphasis on the southern hemisphere in terms of historicity and an imperial framework should be tempered with a contemporary analysis of internal colonialism and late twentieth-century manifestations of expropriative relations within economic neocolonialism. This will enable new formulations of the older concepts and a more sophisticated account of the geography of the global political economy (Corbridge, 1993; Agnew and Corbridge, 1995).

As the developments in the post-Cold War situation have raised new questions, such as the centrality of the state within a highly integrated world economy, so the role of the southern hemisphere within the global political economy also requires serious theoretical reconsideration, as indeed it did at the height of the Cold War. To persist in the view of the southern hemisphere in Cold War terms is to deny many of the rapid changes that have taken place in terms of regional reorganisation, environmental debates, fundamentalism, and social mobilisation: changes arguably as rapid as those witnessed in Eastern Europe, where Northern eyes have tended to be more focused. In Central America, for example, rapid changes have followed the events of the 1989 Panama invasion, the 1990 UNO victory in Nicaragua, the laying down of arms and the renewed 'recontra' threat, and the Arena–FMLN political arrangements in El Salvador.

The need to broaden the base of political geography is a criticism of the failure of northern hemisphere political geographers to engage with the interface of political geography and issues of development in southern hemisphere contexts during the 1970s and 1980s. During the 1990s a vanguard of political geographers have set in motion several stimulating debates to ensure that a high profile for political geography is maintained and that political geography holds its own alongside the theoretical developments in international relations and international political economy (see Agnew, 1992; Dalby, 1990b, 1991; Taylor, 1993a, 1993b; O Tuathail, 1992). However, a vacuum has existed within these arguments and discourses.

In the vacuum is the need for new perspectives on the South, Latin America in this case, and the value of a democratic political geography that seeks to understand the power-space relations of the household as much as international boundary disputes. Democratic political geography would suggest that no spatial scale is more or less significant than another, and that no social group (and their discourses) is more significant than another. Power-space relations operate at all levels and may be analysed at these multiple levels.

By shifting attention away from the preoccupation with the nation-state within political geography, the need to explore other scales of analysis is highlighted. This does not deny that the nation-state is still fundamental to explanations of spatial organisation and how power operates within and between spaces at particular scales. However, the activities of Latin American popular movements and the significant changes within the political organisation of civil society provide strong reasons for extending the remit of political geography into wider society, beyond the confines of the state and its activities. Democratic political geography must look into processes and activities operating 'beyond' the state, in the 'Huallaga Republic' in Peru, for instance, or in the domestic space of the home.

While it could be argued that these spaces lie within national territorial spaces and therefore fall into the realm of the nation-state, the power-space relations established within them cannot be explained via an analysis of the state alone. The power-space relations that operate within households may have little connection with the organisation of the state and its control. Only by examining these 'Other' spaces and the power relations that are established within them can a fuller picture of power, space and their interactions be painted by political geographers. Power sharing is resisted by the close associations of economic élites and political élites. It is not until the social structures of Latin American societies are transformed beyond their present confines and hierarchies that democracy and democratic transitions can be established as a normal state of political affairs on the continent. Ali Abootalebi (1995), in a discussion of democratisation in the South during the 1980s, writes that as long as the state remains the ultimate centre of power, prospects for democracy will remain weak.

New perspectives must be able to reflect all the principal changes in 1990s society that Osvaldo Sunkel (1995, 45) terms, in the Latin American context, the 'demand for democracy', incorporating:

the increasing generalized aspirations and demands for greater participation in the economic realm (ownership, incomes, consumption,
jobs); in the social arena (human rights, social mobility, grass-roots

organizations); in political activity and power (elections, citizenship, participation in decisionmaking); and in cultural life (access to education, information, the media, and cultural goods and services generally). These demands have spread and become particularly strong among former relatively excluded sectors such as women, youth, religious minorities, regional groups, and ethnic minorities and majorities.

Political geography should extend beyond the state in that it should be inclusive of all people, not only 'recognised' citizens such as registered voters. It is with this in mind that an understanding of the role of the informal sector, or those groupings operating beyond the state, should be facilitated, engendering a geopolitics of inclusion, of democratic political geography, rather than a generally exclusionary geopolitics of the nation-state. The informal sector operating outside recognised formal public and private sector operations, within which as much as two-thirds of Latin America's economically active population may be engaged, provides an area of investigation that political geographers must also confront. The qualification of informal sector participants as 'forgotten' by the state, or rather those who direct or usurp its functions, should not thus render them forgotten by the political geographer.

When, in the period after independence, indigenous groups were harried and slaughtered, it was the forces of the newly established states that undertook this task, which suggests that the question of representation and power between indigenous groups and those 'citizens' of the post-independence states must be addressed and reformulated (Vitale, 1992). In this regard the issues of land ownership, human rights and political organisation and participation have been brought to life by groups and key individuals, such as Rigoberta Menchú and Chico Mendes (the Brazilian rubber tapper and environmentalist murdered in 1988). If political geographies of Latin America take account of democratic political geographies beyond the geopolitical norms of statecraft and inter-state relations, or as Simon Dalby (1991) describes it, 'high geopolitics', groups marginalised by state activities and their power relations may be accounted for rather than excluded. In the rapidly evolving, turbulent, transitions of the early 1990s it is these same groups that may provide better clues to the ordering and/or disordering of local, national and international affairs.

Having made the point for a democratic political geography, it is important to add that, within such a political geography, the nation-state still has an important role to play, although not as the 'ultimate centre' (Dicken, 1992; Carnoy, 1993). Richard O'Brien's (1992) arguments that the globalisation of finance marks 'the end of geography' and the nation-state is premature (Martin, 1994; Agnew and Corbridge, 1995). However, he writes of a situation of 'competition among rules' which is more flexible and manageable than the preoccupations of global governance. If anything, the reorganising of the international economy along global lines from the 1970s has highlighted spatial change, hence geography (Thrift, 1989). There is a confusion of the world order picture since the picture, or map, has traditionally relied on representations of territory and 'fixed' nation-states. In the Latin American case, the forces of transboundary neoliberalism and popular movements have made territories more flexible. Different approaches for understanding the processes leading to this flexibility of organisation and reorganisation are now required.

The nation-states predominant in international affairs before the Second World War influenced trade and commerce via foreign policy, often colonial policy. They now find that the tables are turning. The re-articulation of economic space at a global level has led to nation-states responding to their integration within the international division of labour (Sánchez, 1992). Transboundary flows of capital, information and people make such a task increasingly problematic. It is only through an understanding of the interrelatedness of factors of change, by way of new IPE and a grasp of geopolitical economy, that late twentieth-century Latin American development and change can be conceptualised. For example, while bottom-up approaches place people first, centralising them in terms of analysis, an understanding of Latin American development and change cannot be provided by such perspectives alone.

The neoliberalism of the geopolitical economy has had profound effects on power-space relations in Latin America, in many ways stimulating the bottom-up approaches that are now so important in shaping spaces and relations among social groups. Neoliberal ideas have led to a reduction of effective redistributive policies and to the flow of capital out of Latin America rather than within and around the continent. The results can be observed in increased levels of poverty and a reduction in development, if the term is to be assessed by quality of life indicators in preference to macroeconomic indicators. Even with economic data, used within ECLAC's poverty index, 35 per cent of Latin American society was identified as poor and 15 per cent as indigent in 1980. By 1990 these figures had risen to 39 per cent and 18 per cent respectively. Concentration on income per capita levels in terms of economic policy decision-making has obscured the fact that in Brazil, for example, income per capita is higher than in Costa Rica but Brazil has twice the poverty rate of Costa Rica (Helwege, 1995). Geographical issues clearly need to be considered in such debates. Whatever indicators are used, it is clear that economic and social polarisation has been exacerbated in Latin America.

The early 1990s have seen a slowing in the decline of poverty in Latin America, bucking the downward trend of the 1980s. However, closer examination of the geographies and composition of the affected groups reveals further difficulties in terms of who to target and where in terms of development programmes. Clearly, in some areas of Latin America, groups are organising their own survival and development responses. For many

Andean social groups, for example, the drugs industry has succeeded in ameliorating the costs of national restructuring. The urban informal sector also, like the drugs industry, serves to distort regional and national economies as the prices of the factors of production are considerably altered due to the large numbers of people engaged in informal and illegal economic activities (Alvarez, 1995). A shift in the balance of political geography needs to reflect these responses, the failure of Latin American institutional politics and processes to provide basic needs, and the power-space relations exercised within, and in opposition to, neoliberal policy-making.

POLITICAL GEOGRAPHIES FOR THE TWENTY-FIRST CENTURY

In 1980, Abimael Guzmán pronounced that 'The masses are the light of an emergent new world' (Guzmán, 1980). He was speaking at the close of the First Military School of the Shining Path. In his Marxist-Leninist-Maoist terms, the masses had to seize this emergent new world from their oppressors, by force. It is clear that while such revolutionary rhetoric is still important in many political movements within Latin America, it is not now as popular as the strength of mass movements within civil society and communities engaged in grassroots, self-help activities. The emphasis has shifted from political ideologies to social strategies, such as basic needs provision. In this way, by looking at non-élite groups in society, relationships can be made between the organisers and controllers of the state (traditional political geography) and the citizens of the state, with their local and community forms of organisation. In his *Political Geography of Africa*, E.A. Boateng (1978, 5) also emphasises the importance of such a perspective:

> No politically organised area, however small or however parochial it may wish to be, can live in complete isolation; hence the need to extend our field of enquiry in political geography to the relationships between different political entities from the smallest to the largest and the way in which these relationships are influenced by geographical conditions.

The New World (Dis)Order for the masses is characterised by peaceful cooperation and action to meet their basic needs and ameliorate their social conditions. Such activities may be seen by radicals as merely papering over the cracks rather than engaging in changes to the structure of societies and the causes of uneven development, impoverishment and inequalities. However, many Latin Americans have turned to peaceful solutions to change, offering an alternative to the bloody change advocated by Eurasian-based political ideologies and their own Latin American brands of authoritarianism. Central America's 'lost decade' of the 1980s, the Southern Cone authoritarian states from the 1960s to the 1980s, the terrorist wars in Colombia and Peru, and other violent periods in postwar Latin American national histories have led to the search for alternative strategies for initiating change, essentially peaceful ones of direct action from the bottom up, whether they may be soup kitchens, self-help organisations or non-violent demonstrations.

The current fluidity of the global political economy provides a suitable opportunity to reflect on the issue of how geographers are researching, writing and interpreting political geographies of Latin America. Moreover, the fracturing of the bipolar world into a multipolar one from the end of the 1980s has broadened the research agenda. The removal of the dominant superpower players and the nuclear threat from centre stage has given room to rising issues that include concerns for the environment, identity and ethno-territoriality, among others.

The post-Cold War period has been characterised by the recognition of diversity and it is within this context that Latin America must be recognised in all its diversity, not merely within the context of peripheral nation-state players within a Northern-based ideological struggle. The globalisation of information and communication channels has been a principal component of these changes, leading to the increased importance of information supply and consumption in power-space relations. Horacio Godoy (1993, 71) refers to 'thousands of artificial satellites [that] are silent witnesses of the emerging global information society'. If there is indeed a case for such a truly global information society, there still remains an undeniable bias in terms of the information being circulated and its origins, in view of the predominance of the English language and the geography of publication and distribution facilities.

Northern academia is constructed as being of, and is more often than not uncritically accepted as having, a higher status within academic circles. This formulation of academic imperialism has been an important factor in reducing the recognition of Southern agency in terms of change and development within the southern hemisphere. The recent literature emerging on the New World Order underlines this point in that it is predominantly North American and European in origin. It is only through an engagement with critiques of this state of affairs, which are strongest within development theory, that political geography can get to grips with underlying problems relating to knowledge construction and dissemination, and connecting issues of authority and control.

The late twentieth century, following phases of economic growth and recession and various systems of social organisation, control and care, currently reveals crises within global and local spatial arenas: the break-up of the former Soviet Union and Yugoslavia, the crises in Central America, Somalia and Rwanda, and the rise of social movements within grassroots civil society. These changing circumstances have given rise to many new spatial forms with which political geography has to engage (Ossenbrügge, 1983–85). Within this contextual space, there is room for new conceptualisations of the state. If Luis Vitale's (1992) notion of the Latin American state is to be believed, it serves only to articulate the processes of capitalist accumulation of transnational companies. Where do indigenous peoples and those engaged within the informal sector find a place within such a state? Even a broader interpretation of the Latin American state cannot mask its post-independence role of association with external finance and commerce and the integration of the local élites and bourgeoisie (for whom the state provided) with the foreign merchant houses and international capital.

While *dependentistas* would emphasise that little has changed in 150 years, the nature of the *fin de siècle* Latin American state and whom it serves requires consideration. The importance of incorporating groups that might be described as 'beyond' the state in such considerations challenges traditional geopolitical analyses. The informal sector and a recognition of ethno-territorial diversity goes a long way towards an understanding of intra-state fracturing, the areas of state breakdown and alternative arenas or sites of power-space relations. The incorporation of these and similar issues makes the conceptual movement towards a democratic political geography of diversity and inclusion.

In light of the above, development geography provides political geography with useful alternative approaches. In preference to an analysis of the state 'top down', there is much to offer in a 'bottom up' approach whereby the participants within the state, and those 'beyond' the state, take precedence over the institutions of, functions of, and operators of, the state. Within such an approach, the informal sector and indigenous groups become central rather than peripheral, thus challenging, or at least problematising, the traditional concepts of citizenship, representation and the role of the state.

An extension of such a perspective involves an analysis of production and economic structures and activities in addition to the state. The state does not exist in isolation; the government of the day co-opts it to facilitate or block different economic activities and as such dictates a national economic policy. Those interest groups that control the dominant economic activities are most often found in control of the Latin American state through government (democratic or authoritarian); the landowning and agribusiness élites of Central America provide good examples. The focus of geopolitical economy on states, firms, technology and markets can help to delve deeper into power-space issues than a purely state-centric approach. There is no need to balance the relationship of the political and the economic within political economy and in the same vein neither geopolitics nor geoeconomics can be held in higher regard than the other. The two are tightly intertwined, inextricably bound together by the transactions of individuals and institutions at all levels (horizontal and vertical) and through space (local to global).

While this text has referred to Latin American nation-states and social groups within civil society as almost synonymous, it is important to stress that it is as difficult to generalise about Latin American states and societies as it is about Central or South American states and societies. All are different, with contrasting histories and contrasting make-ups. However, the idea of Juan Ferreira Pinho (1994) of a Latin American *estado-continente* (continental-state), reconstituting Bolívar's dream, rather than multiple *estado-naciones* (nation-states) would appear to be playing on the optimism of the times as elements of inter-state 'mutual confidence' are strengthening throughout the continent. There are clear signs that Latin America's regions are establishing themselves within the current ideological interregnum, competing as regional groupings with other regional groupings, among and between MERCOSUR, the Andean Pact, MERCOMUN, NAFTA, APEC, the EU and ASEAN. There are also important lessons of reconciliation, negotiation and inter-state coordination that have led to the pacification of the 1990s and these regional orientations.

States and societies within Latin America are currently undergoing rapid change, re-articulating their internal and external relations in the face of the post-Cold War developments. In his article on the new map of global geopolitical equilibrium, Saul Cohen (1982) ended with the following few words about the South generally that are as valid now as fifteen years ago:

> Finally, in viewing the new and unfolding map of world geopolitical equilibrium, we should be mindful that the North–South dialogue has been cast too simplistically. The South or the Developing World is not a unified entity. It consists of diverse clusters of nations in different regions, with varying potential and at various stages of maturity.

A universalising theoretical orientation for Latin America is unrealistic and, in all likelihood, unhelpful. However, by focusing on nations, or groups of nations, with apparently similar positions within the global order one may attempt to make inroads into responding to specific situations, issues or debates, whether economic, political, social or environmental. What is plain to see is that there is a need for greater awareness and involvement of Latin American political geographies beyond traditional geopolitics. In the post-Cold War transition, a better understanding of the operations of societies and their political economies within the new multipolar global arena must lend itself to a fuller comprehension of not only global interdependence but also of inter- and intra-territorial relations, within and between political communities.

For Cuba under Castro (1959–), Nicaragua under Ortega (1979–90) and Peru under Velasco (1968–75), the relationship with Northern states and companies was considered to be so one-sided that revolution and isolationism proved to be more favourable courses of action. Despite the consequences, drastic measures were weighed up and adopted in pursuit of greater political and economic autonomy. Expropriation from Latin America to the USA, Europe and East Asia on the one hand links countries and on the other polarises them, in the same way that there are linkages and divisions within nations. The multiplicity of power relations that this perpetuates and the ways in which these alter control within societies and through space requires further research and theoretical development. Political geographers should take advantage of the new challenges presented by post-Cold War reorientation and address Latin American issues as central to the global political economy rather than peripheral to it.

Regardless of the political and the economic, it is the issue of power that lies at the heart of issues of nation-state, civil society, firm and market operations. This struggle for power defines the relationship within and between political institutions, civil societies and private organisations, and their control over space at the level of the global and the local – the 'glocal' (Gill, 1992; Swyngedouw, 1992). Susan Strange (1994b) notes that power is the common denominator of the market economy and the state system. Until power is tackled as a conceptual issue, however grey an area it may appear, it may be difficult to get past a superficial understanding of glocal transfers, policy-making, competition and conflict. The unpacking of terms such as 'the exercise of power', 'sources of power', and 'powerlessness' is part of this process.

In his comments on 'Five Centuries of the New World Order' (which he dates from the conquest in terms of American integration into a capitalist system), Heinz Dieterich (1992, 52) refers to Franz Fanon's (1967) critique of development, *The Wretched of the Earth*:

> The global village is *structurally* divided into mansions and ghettos. Getting rid of the ghettos requires getting rid of the systems of the elites that produce them. This is the meaning of the Quincentenary and the only hope for the wretched of the earth.

Until the social structures and power relations embedded within Latin American societies undergo change (reformist, revolutionary or otherwise), which is unlikely to be peaceful within the contemporary neoliberal model of capital accumulation without redistribution, the political geographies of the continent will continue to reflect the dominance of a small élite minority integrated into the global political economy and the subordination of an impoverished majority who fend for themselves, peripheral to the needs of capital and subject to ambiguous codes of human rights and morality. Democratic political geographies should reflect the Janus face of contemporary Latin American power-space relations: economic growth and investment under neoliberalism on one face; poverty and self-help coping strategies on the other.

In 1969, Preston James wrote that (1969, 883) 'There are many Latin Americans who question sincerely whether the change from a horse-and wagon economy to an automobile economy, with its accompanying air pollution, and its effect of undermining social and moral values, really represents progress.' Almost thirty years later, the same questionmarks hang over the transition from the automobile economy to the advanced technology economy, especially when a significant proportion of Latin American society is still firmly entrenched in the horse-and-wagon economy. It is the polarisation of these two modes of existence, high and low technologies (regardless of how appropriate they may be), that reveals Latin America as a continent of striking contrasts.

Power-space relations are complex. Not only the traditional wielders of power, societies' élites, but also those traditionally subordinated by power relations are deserving of the recognition of their political geographies. Political geographers would be unwise to ignore the mobilisation of 'people power' within Latin American civil societies as the new millennium dawns. To enable democratic political geographies to act as critiques of contemporary power geometries within Latin America, there is a need to engage with the constructions and discourses relating to these structures and policies. The ways in which indigenous groups and homosexual groups are constructed by dominant groups in Latin American societies, in which heterosexual relations and socio-economic activities are constructed, and the ways in which spaces are constructed as frontiers, wildernesses, problem areas and self-help communities, as examples, must be reviewed critically in order to examine the evolution of discourses that serve to constrain or empower people living in particular spaces.

One can argue that there are many different considerations that face the geographer writing about the South due to particular historical and political circumstances. For this reason, it is proposed that emphasis within political geography should be placed on broadening its focus. It is only by extending the vision of the field and embracing the numerous and contextually complex issues that face individuals, communities, societies and institutions within the South that political geography can move beyond the restrictions of its 'core' intellectual traditions based on the imperial and neocolonial experiences of hegemonic groups and states. Political geographies have equal weight regardless of location and construction, therefore one has to question and challenge the tradition of Northern predominance and bias in twentieth-century political geography. It is only by moving beyond these constraints that political geography can establish itself as a more representative, democratic sub-discipline. Latin America and its peoples can only gain from such a development.

NOTES

CHAPTER 1: POWER TO THE PEOPLE?

1 The most influential framework is world systems analysis. This framework was established by Immanuel Wallerstein (1974, 1980, 1983), following the work of Fernand Braudel. World systems traces the development of a capitalist world economy from a set of mini-systems, through transitions of empires, to the contemporary world economy of the late twentieth century. Peter Taylor is the most influential political geographer to frame his writings within this analytical meta-theory (Taylor, 1993a). For useful explanations and critiques of world systems, see Shannon (1989) and Chirot and Hall (1982).

2 The classics of this genre include García Márquez's *A Hundred Years of Solitude*, Allende's *Eva Luna*, and Carpentier's *The Lost Steps*.

3 American in this context means North American. The adoption of the term by North Americans, specifically in the USA (United States of *America*), reflects a dimension of the linguistic discourses of power exercised with respect to Central and South America.

4 Osgood Hardy (1914, 509), a contemporary of Bowman, wrote that the Andean Indian, 'has no ambition, perhaps because of the repression of his ancestors in the colonial period, or owing to the effects of constant coca chewing. At any rate, all the Indian wants is barely enough to live on. Higher wages, therefore, offer no inducement to him to work harder, but rather less. As the Indian's food is chiefly parched corn, dried mutton, and potatoes, the production of which requires only a minimum of work, it will be seen that there is little inducement for him to work – consequently he does not.'

5 For a discussion of the ethnocentricity of representations in terms of development, see Hettne (1995), also the exchange in *Area* between James Sidaway (1992, 1993) and Claire Madge (1993).

6 It is also appropriate to acknowledge other claims of 'discovery' and conquest at this point, such as theories of Nordic 'discovery' of the Americas from the North.

7 The significance of 1492 was discussed within academic disciplines, from history to anthropology, and in diverse social and racial groups during the Quincentenary of the 'Discovery' in 1992. The geographical location, race and positionality of people relative to the debate dictate their perspectives to a great extent. For a broad discussion of the event, rather than an ill-conceived notion of a 500-year 'celebration', see the articles in *Political Geography Quarterly* 11:4 (1992). Perhaps a better angle to reflect on 1492–1992 is that of a 500-year holocaust, the

genocide of Amerindian peoples by violence and disease. In the same way that it would be inconceivable that a 500-year anniversary of the Jewish holocaust in Europe during the Second World War would be marked by 'celebrations', one wonders to what extent histories of the Iberian genocide of the Amerindians have failed to inform contemporary public opinion. The history of the conquest was written by the victors but, increasingly, historical accounts and contemporary sources are providing the 'secret' histories that reveal the extent of the atrocities.

8   A distinction may be made between traditional political geography and the 'new' political geography (based on critical geopolitics) which has emerged within the sub-discipline since the late 1980s. By 'traditional' political geography I am referring to the political geography constructed around the nation-state and with its roots in the geopolitical schools of analysis in Europe and the USA.

9   Territories traditionally refer to land-based (Latin: *terra*) geographical spaces that coincide with the spatial extent of a government's jurisdiction. In this way, territory provides the link between space, in all its forms, and politics since territories are recognised delineated spaces over which a state exerts its authority (Gottmann, 1975). Territorial studies as a term is more common in Latin America that spatial studies. This is a result of the influence of traditional state-centric geopolitics. Bertha Becker (1982, 236) notes that: 'Territoriality is the consumption of territory, the living face of power.' John Gold (1982) emphasises the intrinsic relationship between human behaviour and partitioned and delineated spaces derived from ethology, the study of territoriality and social organisation. Territoriality leads to associations of attachment and exclusiveness. The roots of the term territoriality are not only from *terra* (earth, land) but also *terrere* (to warn or frighten off).

10 James Sidaway (1994) highlights the complexities of the conceptualisation of the political geographies of globalisation with a reference to the role of 'cyberspaces'. Cyberspaces provide a spatial dimension that confounds many territorially based conceptualisations of space that still predominate in political geography.

11 Glassner (1996), for instance, suggests a 'power inventory' which would include factors of territory, population, government, economy, circulation (transport and communications), military strength and foreign relations.

12 Massey coins the phrase in a discussion of the power-geometry of space-time compression and how groups and individuals are connected differently in terms of flows and movements through times and spaces. This differs slightly from the way in which the term is used here, reflecting simply the complexities of power relations among groups and individuals at numerous levels or scales.

13 Halford Mackinder, Friedrich Ratzel, Rudolf Kjellén and Alfred Mahan are widely recognised as the 'forefathers' of the discipline. Writing from the late nineteenth century, they developed theoretical frameworks such as the organic state theory (Ratzel, Kjellén), the heartland theory (Mackinder) and the role of sea power (Mahan). These are recognised as having had considerable relevance to geopolitical change during the twentieth century.

14 The new theoretical directions in political geography are discussed at length in a wide-ranging set of articles in a special issue of *Geoforum*. The articles range from international and global perspectives (Taylor, Lovering), to 'minority' political geographies (Kofman, Bell) and future directions (Sidaway, Clarke and Doel).

15 Maturana (1992, 138) writes the following in this regard: 'Once someone asked me – "What is the role of the men in a matriarchal society?", to which I replied, "The same as now. To participate with the woman in the creation of our daily life, without centring the relationship in discourses of conflict, competition,

domination, authority, hierarchy, struggle, control, ownership, security, certainty, obedience and power, but in collaboration, acceptance, co-inspiration, conservation, assistance, confidence, co-habitation, agreement, sharing, beauty and harmony."'

16 Clearly this text also falls into this trap to some extent, failing to give weight to the abundant wealth of German and Portuguese/Brazilian literature relating to Latin American political geographies. As early as 1970, the role of Anglo-American geographers working on Latin America was being reflected upon. Nunley (1970) suggested that the role should be one of training, advice and assistance, not one of replacing national geographers.

17 Shanahan's (1963) geography focuses on physical features and is based on environmental determinism. However, he also offers some insight into contemporary resource management and sustainability debates in an economic development sub-section entitled: 'The Unintelligent Exploitation of Resources without Regard for the Future'.

18 Reboratti (1982) notes that German influence shaped Brazilian geography until the 1950s, and still influences Costa Rican geography. Southern Cone military geopolitics was also heavily influenced by German geographical thinking.

19 In the late 1960s, a more functional approach to the discipline was adopted, incorporating cultural ecology and regional economic growth within the regional and urban planning that dominated Latin American geography at the time (Watters, 1970; Minkel, 1970).

20 Carlos Reboratti (1982, 399) is critical of the form of geographical development within the continent: 'Academies inherited much of the *fin de siècle* pomp of the old geographical institutes, but nothing else.'

21 The construction of the 'Other' was the way in which conquests and imperial control were justified. By constructing other cultural groups as different, physically, intellectually and spiritually, they could be placed at a lower level on a European scale of civilisation. As 'lesser' human beings, European barbarism, in terms of what was required to bring Christianity and civilisation, became acceptable and justified.

22 Cardoso and Faletto's work was published in English in 1972. Cristóbal Kay (1989) contends that Gunder Frank's early translation into English raised his profile in the dependency theory debate since his work was all that was accessible to non-Spanish readers.

CHAPTER 2: FROM PRE-COLUMBIAN TO POST-COLD WAR GEOPOLITICS

1 There are very wide variations in estimates of pre-Columbian populations, see Denevan (1976).

2 These Amerindian representations of space are also evident in the northern Americas. Barbara Belyea (1992, 275) provides a critical account of cartographic interpretations of these representations, contending that while cartographic historians recognise that Amerindian cartography is '... valid and separate, their definition of "map" remains culture-specific'.

3 Within the Vice-Royalty of Peru, the *mita* required one-seventh of the adult male population to work at Potosí for one year out of seven. This ensured a working population of 13,000 miners (Burkholder and Johnston, 1994). To this day, mining workers in Potosí have short life expectancies due to the poor working conditions.

4 At the time of the creation of the Audiences, universities were established in the

Americas, in Santo Domingo (1538), San Marcos de Lima (1551) and Mexico (1553). These were based on the model of the University of Salamanca (Mendoza Morales, 1989).

5   The Captaincy of Venezuela was incorporated into the Vice-Royalty of New Granada.

6   Brazil was organised into twelve spatial divisions separated by horizontal boundaries. Each division was controlled by a donatary-captain who administered the area in return for previous loyalty to the Portuguese crown. Each donatary-captain had hereditary rights to the land and exercised authority over settlement, judicial procedures, production and taxation (Williamson, 1992).

7   By the 1770s, 75% of Portuguese colonial income was generated by Brazil. By 1807, 60% of Portuguese exports were of Brazilian origin (Collier et al., 1992).

8   Since productive structures and orientations changed little, the Latin American situation conforms with Abdel-Fadil's (1989) notion of 'perpetual colonialism' whereby primary export production persists and political élites link with the colonial or 'new' colonial élites.

9   Bolívar favoured republicanism while San Martín favoured constitutional monarchism. Bolívar's republicanism won the day due to the monarchical associations of the Spanish empire.

10 The influence of the *caudillos*, regional political 'bosses' with their own military retinues, resulted in Colombia having thirty-four Presidents, of which twenty-seven were military officers, between 1826–65 (Reid, 1984).

11 Elena Alvarez (1995, 144) notes that the USA 'apparently uses its policy on drugs as a major litmus test in the allocation not only of its foreign aid but also in using its influence in that granted by multilateral organisations'.

12 In 1990, official US economic assistance to Central America totalled US$985 million as opposed to assistance in 1993 of only US$612 million. Official military assistance plummeted from US$113.1 million to US$52.1 million (*Inforpress*, 29 October 1990).

13 For a US perspective on resource and development issues in Latin America and the changing nature of US–Latin American relations over the longer term, see Maguire and Welsh Brown (1986).

14 The Chilean Admiral Merino utilises the term *oceanopolítica*, oceanpolitics, in his book of the same name.

15 Pablo Neruda, the Chilean poet, referred to Chile as 'the thin country'.

CHAPTER 3: LATIN AMERICAN GEOPOLITICAL ECONOMY

1   The term rimland was used by the US geopolitician Spykman to define those areas along the periphery of continental Eurasia. The US domino theory assumed that if one state sympathised with the USSR, neighbouring states would follow suit. The domino theory influenced US involvement in Korea and Vietnam, preventing the spread of communism within East and South-east Asia.

2   For an examination of differing perspectives on the New World (Dis)Order, see the articles emerging from an Association of American Geographers (AAG) workshop on the debate (*Annals of the AAG*, 1992).

3   The 'lost decade' refers to the 1980s which was characterised by the violence perpetuated throughout the isthmus. It is considered a lost decade in that it was a period of crisis and destruction rather than a period of development (however one considers the term). The high numbers of those who lost their lives during

this period bears testament to this; James Dunkerley (1994) puts an estimate at 160,000 lives lost and 2 million people displaced.

4 Whatever the case, it would appear that Jorge Castañeda's (1993, 94) description of US–Latin American relations holds true: he refers to them as the Carmen Miranda syndrome, a theatre of the absurd 'of a typical style and simultaneously baroque and surreal'.

5 Whereas in 1981, exports to the USA from Central America totalled 31.5% with 33.9% of Central American imports from that destination, these figures had risen in 1990 to 40.2% and 43.7% respectively (*Inforpress*, 17 September 1992, based on CEPAL figures, January 1992).

6 Graham Smith (1994, 57) points to the three key processes of global and local spatiopolitical reshaping: processes of globalisation are facilitating greater integration and interconnectedness between states and localities; localised forms of collective action are challenging the nation-state; and these twin challenges raise questions about whether the nation-state is the most appropriate form of 'political community'. In the Latin American context, these three interrelated elements are integral to an understanding of late twentieth-century political geographies.

7 Peter Taylor's (1993c) reflection on geography's engagement with the global emphasises the need to move beyond the limiting constructions of geographies of imperialism and Eurocentricity into the global world of difference, similarities and connections: 'My global Geography explicitly replaces Eurocentric domination with a liberation discourse that converts the monologue into a geographical dialogue ... Geography can become a liberating discipline for teaching an equal dialogue between places.' (194).

8 The term 'globalisation', and degrees thereof, is open to interpretation and is easily contested, see for example Gordon (1988) and Cox (1992).

9 Bretton Woods marked a sea-change in the organisation of international capital transfers. There has been an increase in the number of nation-states, leading to a number of firms being larger in terms of capital reserves than some states, thus confusing the cleaner break of the past between firm and nation-state. Stuart Corbridge (1994, 282–291) provides a chronological overview of the changes in the geoeconomic realm from 1944–92, incorporating the key geoeconomic global players and the financial policies of the nation-states, such as Reaganomics, that facilitated their expansion. Corbridge focuses on the emergence of a tripolar world economy based on the nation-states and firms of the USA, Japan and a united Germany. Susan Strange characterises the period of international financial operations triggered by the OPEC oil price rise of 1973 as 'casino capital' (Strange 1986).

10 Shapiro (1996) notes that the 1990s have seen renewed growth in the Brazilian automobile industry. Mercedes, Toyota, Renault, Hyundai and Asia Motors have all joined the existing major producers: GM, Fiat, Volkswagen and Ford.

11 In terms of the inability, and also lack of desire (in view of its self-help nature), of forms of authority to regulate informal sector businesses, those engaged in such activities work at the local level of the geoeconomic ecumene, beyond the regulation of the nation-state in terms of taxation, and wage and pricing legislation. See for instance Hernando de Soto's (1989) *The Other Path*, and Jim Thomas's (1992) *Informal Economic Activity*.

12 In many cases neoliberalism has replicated the privations of the authoritarian regimes when the state also failed to meet its social obligations of basic needs provisions. During these periods, it was women who organised themselves to provide food and other essentials. Soup kitchens have consistently been the structures put in place to carry this out (see Fisher, 1993).

13 Raúl Prebisch (1965, 205) noted that ISI was necessary to alter the balance of trade relations with the Northern commercial powers: 'Substitution ... is not a doctrinaire whim but a response to objective conditions implicit in the realities of our relations with the centres.'

14 The Andean Group continues to be a weak regional bloc, despite an increase in joint business and industrial ventures emerging from EU–Andean cooperation and incentives. Non-EU countries are also realising more projects within the region (Mikus, 1995). Nevertheless, the varying degrees of instability within the area all mitigate against a strong, creditworthy business environment and healthy economic development.

15 Brazil's difficulties in stabilising its currency, and the economy as a whole, since the mid-1980s have led to new currency denominations and hyperinflation. Argentina followed suit in 1995 with the 'Tango Crash', a financial crisis derived from the banking sector triggered by the 'Tequila Crash' in Mexico that immediately preceded it. With economic instability a structural problem within the group, its coherence will depend on less dramatic economic swings and roundabouts in the future.

16 In its thirty years of existence, the BCIE has loaned US$2,021 million to Central America compared with US$2,017 million by the World Bank in fourteen years (*Inforpress*, 14 March 1991).

17 US assistance via the principal vehicle of US government finance, the International Development Agency (AID), to Latin America as a whole dropped by 36% in 1990, parallel to the increase in flow to Hungary and Poland; the decrease to Central America was 30%. In order to compensate, the USA lowered import taxes on various Central American products but this did not equate with the capital that had been redirected (*Inforpress*, 1 March 1990).

18 The 1929 financial crisis (the Wall Street Crash) initiated a period of international recession. Chile was the worst-hit international economy in terms of loss of foreign exchange earnings and balance of trade due to its dependence on nitrate and copper exports to the USA and the British empire. All Latin American economies suffered and only recovered during the second half of the 1930s.

19 *Caudillos* were, most often, the influential landowners of the post-independence period who were capable of mobilising support through personalistic politics. Guillermo O'Donnell (1973) developed the concept of the bureaucratic-authoritarian state to explain the activities of the authoritarian South American states during the 1970s.

20 According to UNCTAD, transnational corporations in the world's richest countries increased from 7,000 in 1969 to 24,000 in 1994 (*Economist* 30 July 1994, 65).

21 Pablo Escobar of the Medellín drugs cartel, shot dead by Colombian narcotics agents in 1994, was in many ways a 'model' drug baron. His opposition to the state and his appeal to poorer sectors of society provided him with a high degree of popular support. This was consolidated when he ran for local government office in Medellín. As a benefactor, he invested in infrastructure and leisure facilities in particular popular neighbourhoods where his populist politics became very attractive.

CHAPTER 4: THE ORGANISATION OF CONTEMPORARY LATIN AMERICAN SPACE

1 Latin America was transformed in the same way that Mackinder (1904) envisaged that the Russian 'heartland' would be, by a rail network.

2 Regis Debray's (1967) *Revolution in the Revolution?* works through the bases of armed struggle within Latin America, incorporating the idea of *foco*. Actively involved in creating the revolutionary theory behind armed struggle, Debray, a Frenchman, became an important intellectual influence behind the Cuban Revolution.

3 Andreas (1990–91) makes the point that there is a strong Andean history of warrior women. This history stretches from Mama Huaco, the ferocious companion of Inca leader Manco Cápac, in the fifteenth century, to Micaela Bastidas, wife of Tumac Amaru II, in their eighteenth-century struggle against the Spanish.

4 While Perón, Castro and Pinochet have been national political leaders, Rigoberta Menchú (the Guatemalan indigenous rights advocate) and Chico Mendes (the Amazonian environmental activist) have both provided influential voices from within civil society rather than institutional politics.

5 An example is that Chilean television forecasts weather from Arica on its northern border to its sector of Antarctica, raising the territorial significance of the latter within the national consciousness. Easter Island is also included, asserting geopolitical influence and the full extent of Chilean territoriality. The same is true of atlases.

6 The idea of global commons is an extension of Hardin's (1968) thesis on the tragedy of the commons. Hardin's thesis is that shared resources conflict with individual interests and are overexploited. The crisis in international fisheries proves his case. The Antarctic reflects the case of an ongoing agreement to maintain 'a commons' without development, thus not to consider it as a shared resource in economic terms but rather in terms of a shared generation of knowledge regarding the continent.

7 The territorial claimants were Chile, Argentina, Australia, New Zealand, Britain, France and Norway. The non-claimants were the USA, USSR, Japan, Belgium and South Africa. The original twelve signatories have increased to over forty (including states such as Poland, Germany, Brazil, China, Uruguay and India and the consultative members).

8 Chile has based its territorial claims on the 1810 *uti possidetis* since Pedro Sánchez de Hoz was titled governor of *Terra Australis* (the southern land), south of the Magellan Strait to the South Pole, in 1539 and Chilean governors during the colonial period had territorial jurisdiction over the Antarctic (Paravic Valdivia, 1988). The Arturo Prat Chilean base was established in 1947.

9 The years of administration of Porfiriato Díaz from the 1880s (Porfiriato was ousted from power by the 1910 Mexican revolution) were termed the 'Porfiriato'.

10 The terms applied to these settlements vary considerably: *favelas* in Brazil, *poblaciones* in Chile; *ranchos* in Venezuela; *barrios* in Peru; and *villas miserias* in Argentina. English terms have included squatter settlements, spontaneous settlements, and shanty towns. The point to make here is that settlements vary in size, age, service provision, internal organisation, and in the nature of their specific obstacles to development and consolidation. In most cases these settlements are established residential areas with sophisticated internal structures, rather than disorganised 'ghettos' of despair.

11 These problems are most evident within the largest, often primate, cities. Of the 41 'million' megacities within Latin America, six are to be found in Mexico and fourteen in Brazil (Angotti, 1995). Despite this concentration, issues of urban service provision, land-use conflict and poverty characterise Latin American urban spaces, from small towns upwards through the urban hierarchy to primate cities.

12 Teodor Shanin (1976) defines the *campesino* as having four essential and interrelated facets: family agricultural activity as the basis of the social structure; cultivation of land and cereals production as the principal subsistence dietary source; a traditional culture linked to small rural communities; and subordination to external agents (landowning élites).

13 Peru's 1980 Agricultural Promotion and Development Law did not even note the *campesino* community. The legislation restored a free market in agricultural land (similar to Mexico's Article 27) which had been frozen by a previous agrarian reform (Reid, 1984). These approaches to rural development provided fertile opposition for Sendero mobilisation. Another important issue in this regard is the role of *campesino* agricultural strategies in terms of the management of sensitive environments (Ortega, 1983).

14 Galeano (1991) continues by noting that this was also the case in North America during the nineteenth century when the US National Congress usurped Sioux lands with the justification that, 'community property is dangerous to the development of the free enterprise system'.

CHAPTER 5: COMMUNITIES OF POVERTY, BODIES OF POWER

1 These might be defined as the key areas of debate in terms of identity. There are others which are not investigated here but are equally valid and significant, such as physical ability.

2 While much change has taken place in the late twentieth century, one of the earliest examples of the pivotal role of women in Latin America was in the earliest fifteenth-century pre-Columbian-Iberian encounters. 'La Malinche' was an Aztec woman who had been sold to the Mayas and acted as interpreter for *conquistador* Hernán Cortés during his invasion of Mexico and the conquest of the Aztecs. La Malinche was indispensable to Cortés and was also his mistress. The Aztecs gave the name 'Malinche' to Cortés, demonstrating the influence that La Malinche had with the *conquistador*. La Malinche played an important role as a cultural intermediary in this early encounter; however, she is identified historically as having betrayed the Aztecs (Todorov, 1984). From the early fifteenth century to the late twentieth century, Latin American historiographies have effectively silenced accounts of the roles of women. La Malinche and other 'exceptional' women are tips of icebergs in this regard.

3 Lady Elizabeth Repetto (1994), director of a sanctuary for women, notes that in a city as small (in the Latin American context) as Montevideo thirty women died from sexual violence in 1991, every day nine women lose their hearing, suffer detached retinas, broken bones or other injuries, ten receive death threats, three are raped, and six have superficial injuries. Bearing in mind that violence against women often goes unreported due to the power relations of fear that are often associated, these figures can only hint at the scale of incidents of violence against women in Latin America.

4 Bolivia 64.3% versus 48.1%; Guatemala 86.6% versus 53.9%; Mexico 80.6% versus 17.9%; and Peru 79.0% versus 48.7% (Psacharopoulos and Patrinos, 1994, cited in Helwege 1995, 107).

5 Artigas led the independence struggle for Uruguay. O'Higgins did the same in Chile. Both are hailed as the liberators of their respective republics.

6 Umbanda involves the worship of Zumbi, the chief of Palmares, a maroon society of escaped slaves, that survived in the state of Alagoas for a century before the Portuguese wiped them out in 1697 (Burdick, 1992). These historical

roots of Afro-Brazilian culture and history were subordinated within the colonial period and it is only towards the end of the twentieth century that they have been officially acknowledged as part of Brazilian cultural history rather than distinct from it.

7   A further point of contrast was the gendered structuring of many Amerindian societies in that women held equal status to men in community decision-making. The Iroquois, Guaraní and many North American Indian groups were similar in this respect, often with matrilinear and matrilocal forms of family succession and married couple location. This situation contrasted markedly with the patriarchal system brought from Europe and established throughout the Americas.

8   Among numerous examples within former political geographies, an example of this trap is provided by Goblet's (1955, 269) *Political Geography and the World Map*: 'Its [South America's] three great States – Argentina, Brazil and Chile – to which, despite its small extent – one should add Uruguay (politically a South American Switzerland) are white republics, Spanish and Portuguese.'

9   Domitila Barrios de Chungará, a Bolivian woman active in the 'wives' commit-tees connected with the struggles of the Bolivian tin miners against the military dictatorships during the 1970s and 1980s, could be described as one of the forerunners of women's popular political activity. Her book, *Let Me Speak!* (1978), helped to raise the consciousness of other Latin American women in terms of their role in political struggles. Interviews with Rosalina Tuyuc (1994) and Eulalia Yagarí González (1994) reveal the contemporary position of women activists in the struggle for indigenous rights.

10 It should be noted that some Amerindians and blacks fought for the Spanish crown against the independence movement. The regional complexities of the independence struggles led to a diversity of alliances among racial and élite/non-élite groups.

11 Benedict Anderson (1991) notes that Simón Bolívar, the Liberator, had once said that a Negro revolt was a thousand times worse than a Spanish invasion. Another example is Bolívar's 1824 decree granting property rights to individual Indians over communal lands, which enabled creoles to acquire land by purchase or by force. Bolívar also distributed lands among his supporters, exacerbating Amer-indian slave labour. He also managed to reverse San Martín's decrees limiting slavery, and abolishing the *mita* and the tribute system.

12 In the Brazilian case, the outcome of republicanism in Europe was that the monarchy was transported to the colony in 1808 from where Brazil's imperial condition was sustained until the overthrow of Portuguese control in 1889.

13 In 1970, Northern assistance to Latin American NGOs reached $1 billion. By 1990 this figure had increased to $7.2 billion. These funds circulate among Latin America's increasing number of NGOs. Central America alone has 4,000 NGOs in operation (Macdonald, 1995).

GLOSSARY

altiplano	the Andean high plains
autogolpe	a process whereby those in power seize greater control, often by suspending constitutional government
ayllu	an Andean term for family or tribe which also refers to the organisation of these groups in space
barrio	an urban area, often characterised by low income and spontaneous settlement
cabildo	town hall
cacique	local political leader or chief (Mexico)
campesino	peasant agricultural worker
capitanía	a territorial division under which influential *conquistadores* gained new estates within the Americas
caudillo	influential regional political leader in the post-independence period
comadrazgo	godparenthood
conquistadores	conquerors, the generic term for the invading Iberian armed forces of the sixteenth century
criollo	creole, American-born person of Spanish descent
dependentista	a supporter of the dependency theory of Latin American development
desaparecido	disappeared person: person arrested or kidnapped by the military and paramilitary in Argentina (also in Chile, Uruguay and Brazil) principally during the 1970s, and who has never since been seen
ejido	Mexican community land established by the Institutional Revolutionary Party (PRI)
encomienda	system of labour in colonial production
golpe de estado	*coup d'état*, seizure of state control
hacienda	a large farm
indio	an indigenous Amerindian
kuraka	a local Andean chieftain
ladino	Central American for a *mestizo*

latifundio	a large landholding for agribusiness (Brazil)
lebensraum	living space (German)
machismo	behaviour associated with the *macho* male; a gendered power relationship within which men are dominant
maquila/ *maquiladora*	firm involved in export processing, typically textiles and electronic goods
marianismo	behaviour associated with the female embodiment of the virtues of the Madonna
mayorazgo	family inheritance by the eldest in the male line
mestizo	a person of mixed European-Amerindian blood
minifundio	a smallholding for agribusiness (Brazil)
misti	Quechua word for a white outsider
mita	a feudal form of labour service
mulato	a person of mixed negro-European blood
narco	drug trafficker
oidor	member of the audience commissions
peninsular	Iberian-born Spanish person, as opposed to *criollo*
plaza	central square, the centre point of urban areas and location of territorial and administrative functions

REFERENCES

Abdel-Fadil, M. (1989) 'Colonialism' in J. Eatwell, M. Milgate and P. Newman (eds) *Economic Development* (Oxford, Blackwell), 61–67.

Abootalebi, A.R. (1995) 'Democratization in Developing Countries: 1980–1989', *Journal of Developing Areas* 29:4, 507–531.

Agnew, J. (1992) 'The US Position in the World Geopolitical Order after the Cold War', *Professional Geographer* 44:1, 7–10.

Agnew, J. and Corbridge, S. (1989) 'The New Geopolitics: The Dynamics of Geopolitical Disorder' in R.J. Johnston and P.J. Taylor (eds) *A World in Crisis? Geographical Perspectives* (Oxford, Blackwell), second edition, 266–288.

—— (1995) *Mastering Space: Hegemony, Territory and International Political Economy* (London, Routledge).

Agosin, M. and Ffrench-Davis, R. (1995) 'Trade Liberalization and Growth: Recent Experiences in Latin America', *Journal of InterAmerican Studies and World Affairs* 37:3, 9–59.

Agudelo Gómez, M. (1988–89) 'Transcendencia geopolítica y estrategia de las Malvinas', *Boletín de la Sociedad Geográfica de Colombia* 39:122–123, 77–85.

Albert, B. (1983) *South America and the World Economy from Independence to 1930* (London, Macmillan).

—— (1985) 'External Forces and the Transformation of Peruvian Coastal Agriculture, 1880–1930' in C. Abel and C.M. Lewis (eds) *Latin America: Economic Imperialism and the State: The Political Economy of the External Connection from Independence to the Present* (London, Athlone Press), 231–250.

—— (1988) *South America and the First World War: The Impact of the War on Brazil, Argentina, Peru and Chile* (Cambridge, Cambridge University Press).

Alvarez, E.H. (1995) 'Economic Development, Restructuring and the Illicit Drug Sector in Bolivia and Peru: Current Policies', *Journal of InterAmerican Studies and World Affairs* 37:3, 125–149.

Alvarez, S.E. and Escobar, A. (1992) 'Conclusion: Theoretical and Political Horizons of Change in Contemporary Latin American Social Movements' in S.E. Alvarez and A. Escobar (eds) *The Making of Social Movements in Latin America* (Boulder, Westview Press), 317–331.

Amin, S. (1988) *L'Eurocentrisme: Critique d'une Idéologie* (Paris, Anthropos).

Anderson, B. (1991) *Imagined Communities: Reflections on the Origin and Spread of Nationalism* (London, Verso), second edition.

Anderson, T.D. (1984) *Geopolitics of the Caribbean: Ministates in the Wider World* (New York, Praeger).

Andreas, C. (1990–91) 'Women at War', *NACLA: Report on the Americas* 24:4, 20–23.

REFERENCES

Angell, A. and Pollack, B. (1990) 'The Chilean Elections of 1989 and the Politics of the Transition to Democracy', *Bulletin of Latin American Research* 9:1, 1–25.

Angotti, T. (1995) 'The Latin American Metropolis and the Growth of Inequality', *NACLA: Report on the Americas* 28:4, 13–19.

Aravena Ricardi, N. (1988) 'Evolución de Límites y Fronteras entre Chile y Bolivia', *Revista Chilena de Geopolítica* 4:3, 13–33.

Archetti, E., Cammack, P. and Roberts, B. (1986) *Sociology of 'Developing Societies': Latin America* (London, Macmillan).

Armstrong, W. and McGee, T. (1985) *Theatres of Accumulation: Studies in Asian and Latin American Urbanisation* (London, Methuen).

Arnello Romo, M. (1988a) 'El Mar en el Futuro de Chile', *Revista Chilena de Geopolítica*, 5:1, 33–37.

—— (1988b) 'Visión Geopolítica de Chile en el Año 2050', *Revista Chilena de Geopolítica* 4:2, 47–55.

—— (1988c) 'Chile: Territorio y Destino Oceánico', *Chile: Histórico y Geopolítico* 2:4, 52–54.

Arocha Rodríguez, J. (1992) 'Afro-Colombians Denied', *NACLA: Report on the Americas* 25:4, 28–31.

Arriagada, I. (1990) 'Unequal Participation by Women in the Working World', *CEPAL Review* 40, 83–99.

Astudillo Espinosa, C. (1993) 'Aspectos Geográficos de la Amazonia Ecuatoriana', *Revista Geográfica* (Instituto Geográfico Militar, Quito) 32, 83–93.

Baer, W. (1975) 'Import Substitution and Industrialization in Latin America: Experiences and Interpretations', *Latin American Research Review* 7:1, 95–122.

Baerreson, D.W. *et al.* (1965) *Latin American Trade Patterns* (Washington, DC).

Barham, B., Clark, M., Katz, E. and Schurman, R. (1992) 'Nontraditional Agricultural Exports in Latin America', *Latin American Research Review* 27:2, 43–82.

Barkin, D. (1991) 'About Face', *NACLA: Report on the Americas* 24:6, 30–37.

Barraclough, S. (1994) 'The Legacy of Latin American Land Reform', *NACLA: Report on the Americas* 28:3, 16–22.

Barrios de Chungará, D. (1978) *Let Me Speak!* (New York, Monthly Review Press).

—— (1992) 'The Owners of this Land', *Latin American Perspectives* 19:3, 92–96.

Bassin, M. (1987) 'Race Contra Space: The Conflict between German Geopolitik and National Socialism', *Political Geography Quarterly* 6:2, 115–134.

Becker, B. (1982) 'The Political Use of Territory: A Third World Perspective', *International Geographical Union Latin American Regional Conference*, vol. II, *Symposia and Round Tables*, 233–240.

Becker, B.K. and Egler, C.A.G. (1992) *Brazil: A New Regional Power in the World Economy* (Cambridge, Cambridge University Press).

Bell, D.J. (1994) 'In Bed with the State: Political Geography and Sexual Politics', *Geoforum* 25:4, 445–453.

Bergevin, J. (1989) 'A Propos de la Géographie Politique: La Parole est à Friedrich Ratzel', *Cahiers de Géographie du Quebec* 33:88, 59–66.

Berry, B. (1969) 'Review of B.M. Russett's International Regions and the International System', *Geographical Review* 59, 450–451.

Blakemore, H. and Smith, C.T. (eds) (1983) *Latin America: Geographical Perspectives* (London, Methuen), second edition.

Boateng, E.A. (1978) *A Political Geography of Africa* (Cambridge, Cambridge University Press).

Boisier, S. (1988) 'Las Regiones como Espacios Socialmente Construidos', *CEPAL Review* 35, 39–54.

Boletín Antártico Chileno (1984) 'Chile Consolida su Soberanía', 4:1, 59–62.

REFERENCES

Bolívar Castillo, J. (1993) 'Del Pacífico a la red fluvial Amazónico-Atlántica', *Revista Geográfica* (Instituto Geográfico Militar, Quito) 32, 71–79.

Bowman, I. (1914) 'Results of an Expedition to the Central Andes', *Bulletin of the American Geographical Society* 46:3, 161–184.

Brandt Commission (1983) *Common Crisis, North–South: Co-operation for World Recovery* (London, Pan).

Brandt, W. (1980) *North–South: A Programme for Survival* (London, Pan).

Braudel, F. (1994) *A History of Civilisations* (trans. R. Mayne) (New York, The Penguin Press).

Brittin, A. (1995) 'Close Encounters of the Third World Kind: Rigoberta Menchú and Elisabeth Burgos's *Me Llamo Rigoberta Menchú*', *Latin American Perspectives* 22:4, 100–115.

Bromley, R.D.F. and Bromley, R. (1988) *South American Development: A Geographical Introduction* (Cambridge, Cambridge University Press), second edition.

Brown, E. (1996) 'Articulating Opposition in Latin America: The Consolidation of Neoliberalism and the Search for Radical Alternatives', *Political Geography* 15:2, 169–192.

Brzezinski, Z. (1991) 'The Cold War and Its Aftermath', *Foreign Affairs* 71:4, 31–50.

Buchhofer, E. and Guillermo Aguilar, A. (1992) 'Urban Expansion in the Debt Crisis: The Case of Mexico City', *Applied Geography and Development* 39, 48–66.

Burdick, J. (1992) 'Brazil's Black Consciousness Movement', *NACLA: Report on the Americas* 25:3, 23–27.

Burgos-Debray, E. (ed.) (1984) *I, Rigoberta Menchú: An Indian Woman in Guatemala* (London, Verso).

Burkholder, M.A. and Johnson, L.L. (1994) *Colonial Latin America* (New York, Oxford University Press), second edition.

Burt, J-M. and Espejo, C. (1995) 'The Struggles of a Self-Built Community', *NACLA: Report on the Americas* 28:4, 19–25.

Bustamente, F. (1988) *La Antártida y el Pensamiento Geopolítico de las Fuerzas Armadas Chilenas* (Santiago, Facultad Latinoamericana de Ciencias Sociales).

Butterworth, D. and Chance, J.K. (1981) *Latin American Urbanization* (Cambridge, Cambridge University Press).

Buzan, B. (1991) 'New Patterns of Global Security in the Twenty-First Century', *International Affairs* 67:3, 431–451.

Camargo Pérez, G. (1988–89) 'La Geografía en el Poblamiento Precolombiano', *Boletín de la Sociedad Geográfica de Colombia* 39: 122–123, 53–63.

Campbell, B. (1991) 'Beggar Thy Neighbour', *NACLA: Report on the Americas* 24:6, 22–30.

Cano, W. (1990) 'Cuestión Regional y Urbanización en el Desarrollo Económico Brasileño', *Estudios Territoriales* 33, 13–33.

Cardoso, F.H. (1972) 'Dependent Capitalist Development in Latin America', *New Left Review* 74, 83–96.

Cardoso, F.H. and Faletto, E. (1969) *Dependencia y Desarrollo en América Latina: Ensayo de Interpretación Sociológica* (México, Siglo Veintiuno).

—— (1972) *Dependency and Development in Latin America* (Berkeley, University of California Press).

Carnoy, M. (1993) 'Multinationals in a Changing World Economy: Whither the Nation-State?' in M. Carnoy *et al.*, *The New Global Economy in the Information Age* (University Park, Pennsylvania State University Press), 45–97.

Carpenter, T.G. (1991) 'The New World Disorder', *Foreign Policy* 84, 24–39.

Carrier, J. (1991) 'The Crisis of Costa Rica: An Ecological Perspective' in D. Goodman and M. Redclift (eds) *Environment and Development in Latin America*

(Manchester, Manchester University Press).

Carvallo Ortiz, D. (1988) 'Raices de Nuestra Raza', *Chile: Histórico y Geopolítico* 2:3, 64–69.

Castañeda, J.G. (1993a) *La Casa por la Ventana: México y América Latina después de la Guerra Fría* (México, Cal y Arena).

—— (1993b) *Utopia Unarmed: The Latin American Left after the Cold War* (New York, Vintage).

—— (1994) 'Latin America and the End of the Cold War: An Essay in Frustration', in A.F. Lowenthal and G.F. Treverton (eds) *Latin America in a New World* (Boulder, Westview Press), 28–53.

Castells, M. (1977) *The Urban Question: A Marxist Approach* (Cambridge, Massachusetts Institute of Technology Press).

—— (1982) 'Squatters and Politics in Latin America: A Comparative Analysis of Urban Social Movements in Chile, Peru and Mexico', in H. Safa (ed.) *Towards a Political Economy of Urbanisation in Third World Countries* (Delhi, Oxford University Press), 249–293.

Castillo, J. (1986) 'Importancia Geopolítica de la Región Sur del Ecuador', *Revista Geográfica* (Instituto Geográfico Militar, Quito) 23, 7–21.

Cavarozzi, M. (1992) 'Beyond Transitions in Democracy in Latin America', *Journal of Latin American Studies* 24:3, 665–684.

Cavarozzi, M. and Palermo, V. (1995) 'State, Civil Society and Popular Neighbourhood Organizations in Buenos Aires: Key Players in Argentina's Transition to Democracy', in C.A. Reilly (ed.) *New Paths to Democratic Development in Latin America* (Boulder, Lynne Rienner), 29–45.

Caviedes, C.N. and Muller, K.D. (1994) 'Fruticulture and Uneven Development in North East Brazil', *Geographical Review* 84:4, 380–394.

CEPAL (Comisión Económica para América Latina) (1988) *La Evolución Económica del Japón y su Impacto en América Latina* (Santiago, CEPAL).

—— (1995) *Statistical Yearbook for Latin America and the Caribbean, 1994* (Santiago, CEPAL).

Chant, S. (1985) 'Single-Parent Families: Choice or Constraint? The Formation of Female-headed Households in Mexican Shanty Towns', *Development and Change* 16, 635–656.

—— (1991) *Women and Survival in Mexican Cities* (Manchester, Manchester University Press).

Chase-Dunn, C. (1990) 'World-State Formation: Historical Processes and Emergent Necessity', *Political Geography Quarterly* 9:2, 108–130.

Chasteen, J.C. (1996) 'The Prehistory of Samba: Carnival Dancing in Rio de Janeiro, 1840–1917', *Journal of Latin American Studies* 28:1, 29–49.

Chilcote, R.H. (1974) 'A Critical Synthesis of the Dependency Literature', *Latin American Perspectives* 1:1, 4–29.

Child, J. (1985) *Geopolitics and Conflict in South America: Quarrels among Neighbors* (New York, Praeger).

—— (1990) 'The Status of South American Geopolitical Thinking', in G. Pope Atkins (ed.) *South America into the 1990s: Evolving International Relationships in a New Era* (Boulder, Westview Press), 53–87.

Chirot, D. and Hall, T.D. (1982) 'World-System Theory', *Annual Review of Sociology* 8, 81–106.

Chisari, O., Fanelli, J.M. and Frenkel, R. (1996) 'Argentina: Growth Resumption, Sustainability, and Environment', *World Development* 24:2, 227–240.

Chomsky, N. (1993) *Year 501: The Conquest Continues* (London, Verso).

Clark, M.A. (1995) 'Nontraditional Export Promotion in Costa Rica: Sustaining

REFERENCES

Export-led Growth', *Journal of InterAmerican Studies and World Affairs* 37:2, 181–225.

Clarke, D.B. and Doel, M.A. (1994) 'Transpolitical Geography', *Geoforum* 25:4, 505–525.

Clastres, P. (1987) *Society against the State* (New York, Zone Books).

Claval, P. (1986) 'Un Grand Tour parmi les Géographies et les Géographes du Monde Anglo-Americana', *Revista Geográfica de América Central* (Costa Rica) 21–22, 11–23.

Cleary, D. (1991) 'The Greening of the Amazon', in D. Goodman and M. Redclift (eds) *Environment and Development in Latin America: The Politics of Sustainability* (Manchester, Manchester University Press), 116–141.

—— (1993) 'After the Frontier: Problems with Political Economy in the Modern Brazilian Amazon', *Journal of Latin American Studies* 25:2, 331–351.

Cloke, P., Philo, C. and Sadler, D. (1991) *Approaching Human Geography: An Introduction to Contemporary Theoretical Debates* (London, Paul Chapman).

Cockburn, A. and Hecht, S. (1989) *The Fate of the Forest: Developers, Destroyers and Defenders of the Amazon* (London, Verso).

Cohen, S. (1973) *Geography and Politics in a World Divided* (New York, Oxford University Press), second edition.

—— (1982) 'A New Map of Global Geopolitical Equilibrium: A Developmental Approach', *Political Geography Quarterly*, 1:3, 223–241.

—— (1992) 'Policy Prescriptions for the Post Cold War World', *Professional Geographer* 44:1, 13–15.

Collier, S., Skidmore, T.E. and Blakemore, H. (1992) *The Cambridge Encyclopaedia of Latin America and the Caribbean* (Cambridge, Cambridge University Press), second edition.

Comacho, D. (1993) Latin America: A Society in Motion', in P. Wignaraja (ed.) *New Social Movements in the South: Empowering the People* (London, Zed Books), 36–59.

Comisión Económica para América Latina – *See* CEPAL.

Conaghan, C.M. *et al.* (1990) 'Business and the "Boys": The Politics of Neoliberalism in the Central Andes', *Latin American Research Review* 25:2, 3–31.

Coraggio, J.L. (1987) *Territorios en Transición: Una Crítica a la Planificación Regional de América Latina* (Quito, Editorial Centro-Ciudad).

Corbridge, S. (1992) 'Discipline and Punish: The New Right and the Policing of the International Debt Crisis', *Geoforum* 23:3, 285–301.

—— (1993) 'Colonialism, Post-Colonialism, and the Political Geography of the Third World', in P.J. Taylor (ed.) *Political Geography of the Twentieth Century: A Global Analysis* (London, Belhaven), 171–206.

—— (1994) 'Maximising Entropy? New Geopolitical Orders and Internationalization of Business', in G.J. Demko and W.B. Wood (eds) *Reordering the World: Geopolitical Perspectives on the 21st Century* (Boulder, Westview Press), 281–301.

Corkill, D. and Cubitt, D. (1988) *Ecuador: Fragile Democracy* (London, Latin American Bureau).

Cox, K.R. (1992) 'Comment: The Politics of Globalization: A Sceptic's View', *Political Geography* 11:5, 427–430.

Coy, M. (1992) 'Pioneer Front and Urban Development: Social and Economic Differentiation of Pioneer Towns in Northern Mato Grosso (Brazil)', *Applied Geography and Development* 39, 7–30.

Crabtree, J., Duffy, G. and Pearce, J. (1987) *The Great Tin Crash: Bolivia and the World Tin Market* (London, Latin American Bureau).

Cubitt, T. (1995) *Latin American Society* (Harlow, Longman), second edition.

REFERENCES

Dabene, O. (1991) 'Quelle Intégration pour quelle Amérique Centrale dans les Années Quatre-vingt-dix', *Cahiers des Amériques Latines* 12, 165–177.

Dalby, S. (1990a) 'American Security Discourse: The Persistence of Geopolitics', *Political Geography Quarterly* 9:2, 171–188.

—— (1990b) *Creating the Second Cold War: The Discourse of Politics* (London, Pinter).

—— (1991) 'Critical Geopolitics: Discourse, Difference and Dissent', *Environment and Planning D: Society and Space* 9:3, 261–283.

—— (1993) 'Gender and Geopolitics: Discourse, Silence and Violence in the New World Disorder', paper for the annual meeting of the Canadian Association of Geographers, Ottawa, May–June.

Davidson, M. (1991) 'Organizing in the Maquilas', *NACLA: Report on the Americas* 24:6, 20–22.

Dávila, L.R. (1993) 'Rómulo Betancourt and the Development of Venezuelan Nationalism (1930–1945)', *Bulletin of Latin American Research* 12:1, 49–65.

Davis, L.H. (1991) 'Peace and Development in Central America: A Review of Five New Proposals', *The Developing Economies* 29:3, 234–246.

de Blij, H. (1992) 'Political Geography of the Post Cold War World', *Professional Geographer* 44:1, 16–19.

de las Casas, B. (1974) *In Defense of the Indians* (De Kalb, Northern Illinois University).

—— (1995) 'In Defense of the Indians', in O. Starn, C.I. Degregori and R. Kirk (eds) *The Peru Reader: History, Culture, Politics* (Durham, North Carolina, Duke University Press), 107–111.

de Dios Carmona, J. (1987) 'Presencia de O'Higgins en la Institucionalidad Chilena de Hoy', *Chile: Histórico y Geopolítico* 1:1, 7–17.

de Oliver, M. (1993) 'The Hegemonic Cycle and Free Trade: The US and Mexico', *Political Geography*, 12:5, 457–472.

de Soto, H. (1989) *The Other Path* (London, IB Tauris).

Debray, R. (1967) *Revolution in the Revolution?* (Harmondsworth, Penguin).

Degregori, C.I. (1992) 'The Origins and Logic of Shining Path: Two Views – Return to the Past', in D.S. Palmer (ed.) *Shining Path of Peru* (London, C. Hurst), 33–45.

Demko, G. and Wood, W.B. (1994) *Reordering the World Geographical Perspectives on the Twenty-first Century* (Boulder, Westview Press).

Denevan, W.M. (ed.) (1976) *The Native Population of the Americas in 1492* (Madison, University of Wisconsin Press).

—— (1992) 'The Pristine Myth: The Landscape of the Americas in 1492', *Annals of the Association of American Geographers* 82:3, 369–385.

Díaz-Alejandro, C.F. (1965) 'On the Import Intensity of Import Substitution', *Kyklos* 18, 495–511.

Dicken, P. (1992) *Global Shift: The Internationalization of Economic Activity* (London, Paul Chapman).

Dieterich, H. (1992) 'Five Centuries of the New World Order', *Latin American Perspectives* 19:3, 48–53.

Dietz, J.L. (1995) *Latin America's Economic Development: Confronting Crises* (Boulder, Lynne Reinner).

Dodds, K-J. (1993a) 'Geopolitics, Experts and the Making of Foreign Policy', *Area*, 25:1, 70–74.

—— (1993b) 'Geopolitics, Cartography and the State in South America', *Political Geography* 12:4, 361–381.

—— (1993c) 'Geography, Identity and the Creation of the Argentine State', *Bulletin of Latin American Research* 12:3, 311–333.

REFERENCES

—— (1994) 'Geopolitics and Foreign Policy: Recent Developments in Anglo-American Political Geography and International Relations', *Progress in Human Geography* 18:2, 186–208.

Downes, R. (1992) 'Autos over Rails: How US Business Supplanted the British in Brazil, 1910–1928', *Journal of Latin American Studies* 24:3, 551–585.

Doyle, K. (1993) 'Drug War: A Quietly Escalating Failure', *NACLA: Report on the Americas* 26:5, 29–34.

Drysdale, A. and Blake, G.H. (1985) *The Middle East and North Africa: A Political Geography* (Oxford, Oxford University Press).

Dunkerley, J. (1994) *The Pacification of Central America* (London, Verso).

Dunleavy, P. and O'Leary, B. (1987) *State Theory: The Politics of Liberal Democracy* (London, Macmillan).

Eckstein, S. (1989) 'Power and Popular Protest in Latin America', in S. Eckstein (ed.) *Power and Popular Protest: Latin American Social Movements* (Berkeley, University of California Press), 1–61.

ECLAC (Economic Commission for Latin America and the Caribbean) (1995) *Social Panorama, 1994* (Santiago, ECLAC).

—— (1995) *Economic Panorama of Latin America, 1995* (Santiago, ECLAC).

—— (1995) *Economic Survey of Latin America and the Caribbean, 1994–95* (Santiago, ECLAC).

Edelstein, J.C. (1995) 'The Future of Democracy in Cuba', *Latin American Perspectives* 22:4, 7–27.

Escobar, A. (1995) *Encountering Development: The Making and Unmaking of the Third World* (Princeton, Princeton University Press).

Escobar, A. and Alvarez, S. (1992) 'Introduction: Theory and Protest in Latin America', in Escobar, A. and Alvarez, S. (1992) *The Making of Social Movements in Latin America: Identity, Strategy and Democracy* (Boulder, Westview Press), 1–19.

Escolar, M., Quintero Palacios, S. and Reboratti, C. (1993) 'Geographical Identity and Patriotic Representation in Argentina', in D. Hooson (ed.) *Geography and National Identity* (Oxford, Blackwell), 346–367.

Escudé, C. (1988) 'Argentine Territorial Nationalism', *Journal of Latin American Studies* 20:2, 139–165.

Espinosa Fernández de Córdoba, C.R. (1989) 'The Fabrication of Andean Particularism', *Bulletin de l'Institut Français d'Etudes Andines* 18:2, 269–298.

Estaba, R.M. (1991) 'La Decentralización: Un Tema Inicial de Inminente Trascendencia para la Geografía', *Revista Geográfica Venezolana* 32, 21–31.

Evans, P. (1979) *Dependent Development: The Alliance of Multinational, State and Local Capital in Brazil* (Princeton, Princeton University Press).

Eyzaguirre, N. and Valdivia, M. (1989) 'Restricción Externa y Ajuste: Opciones y Políticas en América Latina', in Eyzaguirre, N. and Valdivia, M. (eds) *Políticas Macroeconómicas y Brecha Externa: América Latina en los Años Ochenta* (Santiago, CEPAL), 11–34.

Fals Borda, O. (1992) 'Social Movements and Political Power in Latin America', in A. Escobar and S. Alvarez (eds) *The Making of Social Movements in Latin America: Identity, Strategy and Democracy* (Boulder, Westview Press), 303–317.

Fanon, F. (1967) *The Wretched of the Earth* (Harmondsworth, Penguin).

Farthing, L. and Villegas, C. (1991) 'After the Crash', *NACLA: Report on the Americas* 25:1, 24–28.

Felix, D. (1992) 'Privatizing and Rolling Back the Latin American State', *CEPAL Review* 46, 1–47.

REFERENCES

Fereira Pinho, J.A. (1994) 'Las Medidas de Confianza Mutua en la Realidad Americana', working paper, Centro de Estudios para América Latina: Instituto de Defensa, Cooperación y Seguridad Internacional, Buenos Aires.

Feres, J. and León, A. (1990) 'The Magnitude of Poverty in Latin America', *CEPAL Review* 41, 133–153.

Fernández A., Fernández, L.F. and Barliza Zubiria, R. (1991) 'Análisis Crítico de las Concepciones Neoclásicas y Estructurales del Desarrollo Regional', *Revista Geográfica de América Central* (Costa Rica) 23–24, 99–115.

Fernández-Kelly, M.P. (1983) *For We Are Sold, I and My People: Women and Industry in Mexico's Frontier* (Albany, University of New York Press).

Field, L. (1991) 'Ecuador's Pan-Indian Uprising', *NACLA: Report on the Americas* 25:3, 39–44.

Figueroa, H. (1996) 'In the Name of Fashion: Exploitation in the Garment Industry', *NACLA: Report on the Americas* 29:4, 34–36.

Fisher, J. (1993) *Out of the Shadows: Women's Resistance in Politics in Latin America* (London, Latin America Bureau).

Foley, M.W. (1995) 'Privatising the Countryside: The Mexican Peasant Reform Movement and Neoliberal Reform', *Latin American Perspectives* 22:1, 59–77.

Ford, R.E. and Hudson, V.M. (1992) 'The USA and Latin Amèrica at the End of the Columbian Age: How America "Cut the Atlantic Apron String" in 1992', *Third World Quarterly* 13:3, 441–463.

Foucault, M. (1980) *Power/Knowledge* (New York, Pantheon Books).

Foweraker, J. (1981) *The Struggle for Land: A Political Economy of the Pioneer Frontier in Brazil from 1930 to the Present Day* (Cambridge, Cambridge University Press).

Foweraker, J. and Craig, A. (1990) *Popular Movements and Political Change in Mexico* (Colorado, Lynne Rienner).

Frank, A.G. (1966) 'The Development of Underdevelopment', *Monthly Review* 18:4.

—— (1969) *Sociology of Development and Underdevelopment* (Ormskirk, Causeway Press).

—— (1992) 'Latin American Development Theories Revisited: A Participant Review', *Latin American Perspectives* 19:2, 125–139.

Frazier, L.J. and Scarpaci, J.L. (1993) 'State Terror: Ideology, Protest and the Gendering of Landscapes', *Progress in Human Geography* 17:1, 1–21.

Gajardo Villarroel, E. (1983) 'Chile, el Tratado Antártico y Su Sistema', *Boletín Antártico Chileno* 3:2, 7–15.

Galeano, E. (1991) 'The Blue Tiger and the Promised Land', *NACLA: Report on the Americas* 24:5, 13–18.

Gamble, A. (1993) 'Shaping a New World Order', *Government and Opposition* 28:3, 325–339.

Gangas Geisse, M. and Santis Arenas, H. (1994) 'The Bolivian Maritime Aspiration to Chilean Political Space: Towards a Non-Territorial Solution', in P.O. Girot (ed.) *The Americas: World Boundaries*, vol. 4 (London, Routledge), 170–189.

García Covarrubias, J. (1985) 'Análisis Geopolítico del Desarrollo del Núcleo Vital de Chile', *Revista Chilena de Geopolítica* 2:1, 23–40.

García de León, A. (1992) 'Chiapas and the Mexican Crisis', *NACLA: Report on the Americas* 29:1, 10–13.

García Monge, M. (1990–91) 'Antártica del Siglo XXI: Un Sistema de Cooperación Ejemplar, Tarea para las Nuevas Generaciones', *Boletín Antártico Chileno* 10:2, 27–41.

García Negrete, J. (1993) 'Visión Geopolítica de la Faja Fronteriza en la Región

REFERENCES

Amazónica Ecuador-Colombia', *Revista Geográfica* (Instituto Geográfico Militar, Quito), 32, 59–71.

Gardy, A. (1992) 'Mexico: A Grassroots Challenge', *NACLA: Report on the Americas* 26:3, 8–12.

Gatehouse, M. and Reyes, M.A. (1987) *Soft Drink, Hard Labour: Guatemalan Workers Take On Coca-Cola* (London, Latin America Bureau).

Gay, R. (1990) 'Neighbourhood Associations and Political Change in Rio de Janeiro', *Latin American Research Review* 25:1, 102–119.

George, J. (1993) 'Of Incarceration and Closure: Neo-Realism and the New/Old World Order', *Millennium* 22:2, 197–234.

George, S. and Sabelli, F. (1994) *Faith and Credit: The World Bank's Secular Empire* (London, Penguin).

Gereffi, G. and Hempel, L. (1996) 'Latin America in the Global Economy: Running Faster to Stay in Place', *NACLA: Report on the Americas* 29:4, 18–28.

Ghisolfo Araya, F. (1985) 'Influjo Geopolítico de la Isla de Pascua', *Revista Chilena de Geopolítica* 2:1, 40–51.

—— (1987) 'Bolivia y el Mar: Reflexiones Geopolíticas', *Chile: Histórico y Geopolítico* 1:1, 42–47.

—— (1988) 'El Norte Grande: Apuntes para un Análisis Geopolítico', *Revista Chilena de Geopolítica* 4:2, 55–59.

Giblin, B. (1985) 'Hérodote, une géographie géopolitique', *Cahiers de Géographie du Québec* 29:77, 283–295.

Gierhake, K. (1993) 'Regional Development Planning in Marginalized Andean Latin America: Examples from Bolivia, Colombia and Peru', *Applied Geography and Development* 42, 7–25.

Gilbert, A. (1992) *Cities, Poverty and Development: Urbanization in the Third World* (Oxford, Oxford University Press), second edition.

—— (1994) *The Latin American City* (London, Latin America Bureau).

Gilbert, A. and Ward, P. (1985) *Housing, the State and the Poor* (Cambridge, Cambridge University Press).

Gill, S. (1992) 'Economic Globalization and the Internationalization of Authority: Limits and Contradictions', *Geoforum* 23:2, 269–283.

Gill, S. and Law, D. (1988) *The Global Political Economy* (Baltimore, Johns Hopkins Press).

Gills, B. and Rocamora, J. (1992) 'Low Intensity Democracy', *Third World Quarterly* 13:3, 501–525.

Girot, P.O. (1994) 'The Interoceanic Canal and Boundaries in Central America: The Case of the San Juan River', in P.O. Girot (ed.) *The Americas, World Boundaries*, vol. 4 (London, Routledge), 84–113.

Glassner, M.I. (1996) *Political Geography* (New York, Wiley), second edition.

Goblet, Y.M. (1955) *Political Geography and the World Map* (London, George Philip).

Godoy, H.H. (1993) 'El Proceso de Integración de América Latina y la Nueva Realidad Mundial', in H. Casanova (ed.) *América Latina: De la Marginalidad a la Inserción Internacional* (Santiago, Centro de Investigaciones y Promoción Iberoamerica-Europa), 57–72.

Gold, J.R. (1982) 'Territoriality and Human Spatial Behaviour', *Progress in Human Geography* 6:1, 44–68.

Gómez Insausti, R. (1988) 'La Región Metropolitana de Buenos Aires: Una Desproporcionada Concentración', in J.A. Roccatagliata (ed.) *La Argentina: Geografía General y los Marcos Regionales* (Buenos Aires, Editorial Planeta), 443–469.

REFERENCES

Gonzáles, J.E. (1992) 'Guerillas and Coca in the Upper Huallaga Valley', in D.S. Palmer (ed.) *Shining Path of Peru* (London, C. Hurst), 105–127.

Goodman, D. and Hall, A. (1990) *The Future of Amazonia: Destruction or Sustainable Development?* (London, Macmillan).

Gordon, D. (1988) 'The Global Economy: New Edifice or Crumbling Foundations', *New Left Review* 168, 24–65.

Gorostiaga, X. (1990) *La Triangulación Centroamérica – México – EUA ¿Una Oportunidad para el Desarrollo y la Paz?* (San José, DEI).

Gottmann, J. (1973) *The Significance of Territory* (Charlottesville, Virginia University Press).

—— (1975) 'The Evolution of the Concept of Territory', *Social Science* 14:3, 29–47.

—— (1982) 'The Basic Problem of Political Geography: The Organisation of Space and the Search for Stability', *Tijdschrift voor Economische en Sociale Geografie* 73:6, 340–349.

Green, D. (1991) *Faces of Latin America* (London, Latin America Bureau).

—— (1995) *Silent Revolution* (London, Latin America Bureau).

—— (1996) 'Latin America: Neoliberal Failure and the Search for Alternatives', *Third World Quarterly* 17:1, 109–122.

Green, J. and Asis, E. (1993) 'Gays and Lesbians: The Closet Door Swings Open', *NACLA: Report on the Americas* 26:4, 4–7.

Greenhill, R.G. (1977) 'Merchants and the Latin American Trades: An Introduction', in D.C.M. Platt (ed.) *Business Imperialism 1840–1930: An Inquiry Based on the British Experience in Latin America* (Oxford, Clarendon Press), 159–197.

Griffith-Jones, S. and Stallings, B. (1995) 'New Global Financial Trends: Implications for Development', *Journal of InterAmerican Studies and World Affairs* 37:3, 9–59.

Griffith-Jones, S. and Sunkel, O. (1986) *Debt and Development Crises in Latin America: The End of an Illusion* (Oxford, Clarendon Press).

Ground, R.L. (1988) 'The Genesis of Import Substitution in Latin America', *CEPAL Review* 36, 179–202.

Grugel, J. (1995) *Politics and Development in the Caribbean Basin* (London, Macmillan).

Guimarâes, R. (1989) 'The Ecopolitics of Development in Brazil', *CEPAL Review* 38, 89–103.

Gunson, P., Thompson, P. and Chamberlain, G. (1989) *The Dictionary of Contemporary Politics of South America* (London, Routledge).

—— (1990) *The Dictionary of Contemporary Politics of Central America* (London, Routledge).

Gutiérrez, M.A. (1993) 'La Iniciativa de las Américas, un Marco para la Integración y Reinserción Internacional de América Latina', in H. Casanueva (ed.) *América Latina: De la Marginalidad a la Inserción Internacional* (Santiago, Centro de Investigaciones y Promoción Iberoamerica-Europa), 73–90.

Gutiérrez, R. (1981–82) 'Poder Local y Desarrollo Rural en Panamá: El Ejemplo del Distrito de Guarare', *Revista Geográfica de América Central* (Costa Rica) 15–16, 77–95.

Guttman, W. (1990) 'The Political Economy of South American Debt', in G. Pope Atkins (ed.) *South America into the 1990s: Evolving International Relationships in a New Era* (Boulder, Westview Press), 161–187.

Guzmán, A. (1980) 'We are the Initiators' in O. Starn, C.I. Degregori and R. Kirk (eds) (1995) *The Peru Reader: History, Culture, Politics* (Durham, North Carolina, Duke University Press), 310–316.

REFERENCES

Gwynne, R. (1985) *Industrialisation and Urbanisation in Latin America* (London, Croom Helm).

—— (1993) 'Non-Traditional Export Growth and Economic Development: The Chilean Forestry Sector since 1974', *Bulletin of Latin American Research* 12:2, 147–171.

—— (1994) 'Regional Integration in Latin America: The Revival of a Concept?' in R. Gibb and W. Michalak (eds) *Continental Trading Blocs: The Growth of Regionalism in the World Economy* (New York, Wiley), 189–208.

Hall, A. (1993) 'Non-Governmental Organisations and Development in Brazil under Dictatorshịp and Democracy', in C. Abel and C. Lewis (eds) *Welfare, Equity and Development in Latin America* (London, Macmillan), 421–437.

Handy, J. (1992) 'Guatemala: A Tenacious Despotism', *NACLA: Report on the Americas* 26:3, 31–38.

Hardin, G. (1968) 'The Tragedy of the Commons', *Science* 162, 1243–1248.

Hardy, O. (1914) 'Cuzco and Apurimac: A Report on the Commercial and Industrial Outlook of South Central Peru', *Bulletin of the American Geographical Society* 46:7, 500–512.

Hargreaves, C. (1992) *Snowfields: The War on Cocaine in the Andes* (London, Zed Books).

Harley, J.B. (1992a) 'Rereading the Maps of the Columbian Encounter', *Annals of the Association of American Geographers* 82:3, 522–542.

—— (1992b) 'Deconstructing the Map', in T. Barnes and J. Duncan (eds) *Writing Worlds: Discourse, Text and Metaphor in the Representation of Landscape* (London, Routledge), 231–248.

Hart, J. (1976) 'Three Approaches to the Measurement of Power in International Relations', *International Organization* 30, 289–305.

Hecht, S. (1985) 'Environment, Development and Politics: Capital Accumulation and the Livestock Sector in Eastern Amazonia', *World Development* 13:6, 663–684.

Hechter, M. (1975) *Internal Colonialism* (London, Routledge & Kegan Paul).

Helwege, A. (1995) 'Poverty in Latin America: Back to the Abyss', *Journal of InterAmerican Studies and World Affairs* 37:3, 99–125.

Hemming, J. (1970) *The Conquest of the Incas* (London, Harvester).

—— (1978) *Red Gold: The Conquest of the Brazilian Indians, 1500–1760* (Cambridge, Harvard University Press).

Hennessy, A. (1978) *The Frontier in Latin American History* (London, Edward Arnold).

Henshall, J. and Momsen, R.P. (1974) *A Geography of Brazilian Development* (London, Bell & Hyman).

Henwood, D. (1996) 'The Free Flow of Money', *NACLA: Report on the Americas* 29:4, 11–18.

Hepple, L. (1986) 'Geopolitics, Generals and the State in Latin America', *Political Geography Quarterly* 5 (supplement), S79–S90.

—— (1992) 'Metaphor, Geopolitical Discourse and the Military in South America', in T. Barnes and J. Duncan (eds) *Writing Worlds: Discourse, Text and Metaphor in the Representation of Landscape* (London, Routledge), 136–154.

Hernández, E. (1981–82) 'Relaciones Ciudad-Campo en América Latina: El Caso de Venezuela', *Revista Geográfica de América Central* (Costa Rica) 15–16, 163–174.

Hernández, L. (1994) 'The New Mayan War', *NACLA: Report on the Americas* 27:5, 6–10.

Herzog, L. (ed.) (1992) *Changing Boundaries in the Americas: New Perspectives on the US–Mexican, Central American, and South American Borders* (San Diego, University of California).

REFERENCES

Hettne, B. (1995) *Development Theory and the Three Worlds* (London, Longman), second edition.

Hidrovo Penaherrera, J. (1985) 'Galápagos en la Soberanía Ecuatoriana', *Revista Geográfica* (Instituto Geográfico Militar, Quito) 22, 55–61.

Hocquenghem, A-M. (1991) 'Frontera entre "Areas Culturales" Nor y Centroandinas en los Valles y la Costa del Extremo Norte Peruano', *Bulletin de L'Institut Français d'Etudes Andines* 20:2, 309–348.

Hojman, D. (1990) 'Chile after Pinochet: Aylwin's Christian Democrat Economic Policies for the 1990s', *Bulletin of Latin American Research* 9:1, 25–49.

Holden, R.H. (1993) 'The Real Diplomacy of Violence: United States Military Power in Central America, 1950–1990', *International History Review* 15:2, 282–322.

Holm, H-H. and Sørensen, G. (eds) (1995) *Whose World Order? Uneven Globalization and the End of the Cold War* (Boulder, Westview Press).

Honeywell, M. and Pearce, J. (1982) *Falklands: Whose Crisis?* (London, Latin America Bureau).

Hopkins, T. and Wallerstein, I. (eds) (1982) *World Systems Analysis: Theory and Methodology* (Beverly Hills, Sage Publications).

Horowitz, P. and Sklar, H. (1982) 'Surprises in the South Atlantic', *NACLA: Report on the Americas* 16:3, 2–9.

Huntington, S.P. (1968) *Political Order in Changing Societies* (New Haven, Yale University Press).

Hurrell, A. (1992) 'Latin America in the New World Order: A Regional Bloc of the Americas', *International Affairs* 68:1, 121–139.

Inoguchi, T. (1995) 'Dialectics of World Order: A View from Pacific Asia', in H-H. Holm and G. Sørensen (eds) *Whose World Order? Uneven Globalization and the End of the Cold War* (Boulder, Westview Press), 119–137.

Instituto de Ecología Política (IEP) (1994) *El Tigre sin Selva* (Santiago, IEP).

James, P.E. (1969) *Latin America* (New York, Odyssey Press), fourth edition.

Jenkins, R. (1982) 'European Transnational Corporations in Latin America', in J. Pearce (ed.) *The European Challenge: Europe's New Role in Latin America* (London, Latin America Bureau), 130–146.

—— (1991) 'Lessons from the Gang: Are There Lessons for Latin America from East Asia?', *Bulletin of Latin American Research* 10:1, 37–54.

Johnson, H.B. (1987) 'Portuguese Settlement', in L. Bethell (ed.) *Colonial Brazil* (Cambridge, Cambridge University Press), 1–39.

Johnston, R.J. (1989) 'The State, Political Geography, and Geography', in R. Peet and N. Thrift (eds) *New Models in Geography*, vol. I (London, Unwin Hyman), 292–310.

Katz, F. (1978) 'A Comparison of Some Aspects of the Evolution of Cuzco and Tenochtitlán', in R. Schaedel, J.E. Hardoy and M. Scott Kinzer (eds) *Urbanization in the Americas from Its Beginnings to the Present* (The Hague, Mouton), 203–215.

Kay, C. (1989) *Latin American Theories of Development and Underdevelopment* (London, Routledge).

Kelly, P. and Child, J. (eds) (1988) *Geopolitics of the Southern Cone and Antarctica* (Boulder, Lynne Rienner).

Kennedy, P. (1987) *Rise and Fall of the Great Powers: Economic Change and Military Power from 1500 to 2000* (London, Unwin Hyman).

—— (1993) *Preparing for the Twenty-first Century* (London, Harper Collins).

Kindleberger, C. (1984) 'The 1929 World Depression in Latin America – from the Outside', in R. Thorp (ed.) *Latin America in the 1930s: The Role of the Periphery in the World Crisis* (London, Macmillan), 315–329.

REFERENCES

——— (1986) *The World in Depression, 1929–1939* (Berkeley, California, UCLA), second edition.

Knippers Black, J. (1993) 'Elections and Other Trivial Pursuits: Latin America and the New World Order', *Third World Quarterly* 14:3, 545–555.

Kofman, E. (1994) 'Unfinished Agendas: Acting upon Minority Voices of the Past Decade', *Geoforum* 25:4, 429–445.

Kofman, E. and Peake, L. (1990) 'Into the 1990s: A Gendered Agenda for Political Geography', *Political Geography Quarterly* 9:4, 313–336.

Kohlhepp, G. (1989) 'A Challenge to Science and Regional Development Policy: Reflections on the Future Development of Amazonia', *Applied Geography and Development* 33, 52–68.

Koning, H. (1991) *Columbus: His Enterprise* (London, Latin America Bureau).

Kopinak, K. (1995) 'Gender as a Vehicle for the Subordination of Women Maquiladora Workers in Mexico', *Latin American Perspectives* 22:1, 30–49.

Küppers, G. (ed.) (1994) *Compañeras: Voices from the Latin American Women's Movement* (London, Latin America Bureau).

Lacoste, Y. (1976) *La Géographie, Ça Sert d'abord à Faire la Guerre* (Paris, Maspero).

Lara, C. (1991) 'Contribución Crítica sobre la Interpretación Jurídica de la Declaración de Santiago', *Revista Geográfica* (Instituto Geográfico Militar, Quito) 29, 39–53.

Latin America Bureau (1980) *Paraguay: Power Game* (London, Latin America Bureau).

Lazaro Araujo, L. (1993) 'Planificación Económica, Desarrollo Regional y Ordenación del Territorio', *Cuadernos de Ordenación del Territorio* 2:5, 3–6.

Leftwich, A. (1993) 'Governance, Democracy and Development in the Third World', *Third World Quarterly* 14:3, 605–624.

Leonelli Petazzoni, M. and Benito González, F. (1991) 'Modelos Espaciales Integrados en el Diagnóstico Territorial de Nicaragua', *Estudios Territoriales* 36, 147–162.

Levine, D.H. (1993) 'Constructing Culture and Power', in D.H. Levine (ed.) *Constructing Culture and Power in Latin America* (Ann Arbor, University of Michigan Press), 1–41.

Lewis, C.M. (1983) *British Railways in Argentina, 1857–1914: A Case Study of Foreign Investment* (London, Institute of Latin American Studies).

Lewis, C. and Pile, S. (1996) 'Women, Body, Space: Rio Carnival and the Politics of Performance', *Gender, Place and Culture* 3:1, 23–41.

Leyshon, A. (1992) 'The Transformation of Regulatory Order: Regulating the Global Economy and Environment', *Geoforum* 23:3, 249–267.

Lipietz, A. (1987) *Mirages and Miracles: The Crises of Global Fordism* (London, Verso).

Lipton, M. (1977) *Why Poor People Stay Poor: A Study of Urban Bias in World Development* (London, Temple Smith).

Little, W. (1987) 'Paxamericana: International Conflict in Latin America', *International Affairs* 63:3, 589–601.

Livingstone, D.N. (1992) *The Geographical Tradition: Episodes in the History of Contested Enterprise* (Oxford, Blackwell).

Lockhart J. and Schwartz S.B. (1983) *Early Latin America: A History of Colonial Spanish America and Brazil* (Cambridge, Cambridge University Press).

López, C. and Pollack, M. (1989) 'The Incorporation of Women in Development Policies', *CEPAL Review* 39, 37–47.

López de Mesa, L. (1988) 'Simón Bolívar y la Cultura Iberoamericana', *Chile: Histórico y Geopolítico* 2:3, 22–29.

REFERENCES

Lovering, J. (1994) 'The Production and Consumption of the "Means of Violence": Implications of the Reconfiguration of the State, Economic Internationalisation, and the End of the Cold War', *Geoforum* 25:4, 505–525.

Lowenthal, A.F. and Treverton, G.F. (1994) *Latin America in a New World* (Boulder, Westview Press).

Lustig, N. (1995) 'Equity and Development', in J.L. Dietz (ed.) *Latin America's Economic Development: Confronting Crisis* (Boulder, Lynne Rienner), second edition, 55–69.

Luzón J.L. (1987–88) 'El Estado de la Geografía en Cuba', *Revista de Geografía* (Universitat de Barcelona), 21–22, 103–109.

Luzón, J.L., Baila, J. and Sardana, F. (1990) 'Sociedad y Espacio en La Habana de 1877: Un Ensayo de Geografía Urbana Histórica', *Revista de Geografía* (Universitat de Barcelona), 24, 69–87.

Lynch, J. (1973) *The Spanish American Revolutions, 1808–1826* (New York, W.W. Norton).

Macario, S. (1964) 'Protectionism and Industrialisation in Latin America', *Economic Bulletin for Latin America* 9:1.

Macdonald, L. (1995) 'A Mixed Blessing: The NGO Boom in Latin America', *NACLA: Report on the Americas* 28:5, 30–3.

Mackinder, H.J. (1904) 'The Geographical Pivot of History', *Geographical Journal* 23:4, 421–440.

MacRae, E. (1992) 'Homosexual Identities in Transitional Brazilian Politics', in A. Escobar and S. Alvarez (eds) *The Making of Social Movements in Latin America: Identity, Strategy and Democracy* (Boulder, Westview Press), 185–207.

Madge, C. (1993) 'Boundary Disputes: Comments on Sidaway (1992)', *Area* 25:3, 294–299.

Maguire, A. and Welsh Brown, J. (eds) (1986) *Bordering on Trouble: Resources and Politics in Latin America* (Bethesda, Adler & Adler).

Mainwaring, S. (1995) 'Democracy in Brazil and the Southern Cone: Achievements and Problems', *Journal of InterAmerican Studies and World Affairs* 37:1, 113–181.

Malloy, J.M. (1989) 'Policy Analysts, Public Policy and Regime Structure in Latin America', *Governance* 2:3, 315–338.

—— (1993) 'Statecraft, Social Policy, and Governance in Latin America' *Governance*, 6:2, 220–274.

María Miro, L. (1985) 'Génesis de la Soberanía Territorial: Un Intento de Exegesis Histórico-geográfico', *Anales de la Academia Nacional de Geografía* (Argentina) 9, 143–159.

Marichal, C. (1989) *A Century of Debt Crises in Latin America: From Independence to the Great Depression, 1820–1930* (Princeton, Princeton University Press).

Martin, R. (1994) 'Stateless Monies, Global Financial Integration and National Economic Autonomy: The End of Geography?' in S. Corbridge, R. Martin and N. Thift (eds) *Money, Power and Space* (Oxford, Blackwell), 253–279.

Martínez Baeza, S. (1988) 'El Mar, Patrimonio Cultural de la Nación Chilena', *Revista Chilena de Geopolítica* 4:3, 59–63.

Massey, D. (1995) 'Power-geometry and a Progressive Sense of Place', in J. Bird, B. Curtis, T. Putnam, G. Robertson and L. Tickner (eds) *Mapping the Futures: Local Cultures, Global Change* (London, Routledge), 59–70.

Mateo Martinic, B. (1981) 'Estrecho de Magallanes, Territorio Marítimo Chileno', *Anales del Instituto de la Patagonia* 12, 7–31.

Matthews, R.P. (1984) *Regional Atlas of Latin America* (Portsmouth Polytechnic, Department of Geography).

218

REFERENCES

Maturana, H. (1992) *El Sentido de lo Humano* (Santiago, Editorial Universitaria), fourth edition.

Mauro, F. (1978) 'Urban Preeminence and the Urban System in Colonial America', in R. Schaedel, J.E. Hardoy and M. Scott Kinzer (eds) *Urbanization in the Americas from Its Beginnings to the Present* (The Hague, Mouton), 249–269.

McClintock, C. (1992) 'Theories of Revolution and the Case of Peru', in D.S. Palmer (ed.) *Shining Path of Peru* (London, C. Hurst), 225–241.

McCoy, J.L. and Smith, W.C. (1995) 'Democratic Disequilibrium in Venezuela', *Journal of InterAmerican Studies and World Affairs* 37:2, 113–181.

Meller, P. (1989) 'Revisión de los Enfoques Teóricos sobre Ajuste Externo y su Relevancia para América Latina', in N. Eyzaguirre and M. Valdivia (eds) *Políticas Macroeconómicas y Brecha Externa: América Latina en los Años Ochenta* (Santiago, CEPAL), 35–82.

Meller, P., O'Ryan, R. and Solimano, A. (1996) 'Growth, Equity, and the Environment in Chile: Issues and Evidence', *World Development* 24:2, 255–272.

Menchú, R. (1992) 'The Quincentenary, a Question of Class, Not Race', *Latin American Perspectives*, 19:3, 96–101.

Méndez, R. (1986) 'El Resurgir de la Geografía Política', in A. García Ballesteros (ed.) *Teoría y Práctica de la Geografía* (Madrid, Universidad de Alhambra).

Mendoza Morales, A. (1988–89) 'Evolución Histórica de las Divisiones Político-Administrativas de Colombia desde 1509 hasta Hoy', *Boletín de la Sociedad Geográfica de Colombia* 39:122–123, 11–41.

Merino Castro, J.T. (1987) 'La Antártica y el Futuro de Chile', *Boletín Antártico Chileno* 7:1, 27–41.

Mignolo, W.D. (1994) 'The Movable Center: Geographical Discourses and Territoriality during the Expansion of the Spanish Empire', in F.J. Cevallos-Candau *et al.*, *Coded Encounters: Writing, Gender, and Ethnicity in Colonial Latin America* (Amherst, University of Massachusetts Press), 15–46.

Mikus, W. (1995) 'International Industrial Cooperation: Evaluation of Andean Pact Countries', *Tijdschrift voor Economische en Sociale Geografie* 86:4, 357–368.

Milia, J.G. (1982) 'La Usurpación de las Malvinas, Islas Australes y Antártida en el Contexto de la Penetración Británica en el Atlántico Sur', *Boletín de Estudios Geográficos* (Universidad Nacional de Cuyo) 21:80, 105–21.

—— (1990) 'Avances Conceptuales y Metodológicos de la Geografía Política en las Ultimas Décadas', *Boletín de Estudios Geográficos* (Universidad Nacional de Cuyo) 24:86, 67–81.

Miller, M. (1991) *Debt and the Environment: Converging Crisis* (New York, UN Publications).

Miller, R. (1993) *Britain and Latin America in the Nineteenth and Twentieth Centuries* (London, Longman).

Minkel, C. (1970) 'Geography in Central America: Its Status and Prospects', *East Lakes Geographer* 6, 63–74.

Mittelman, J.H. (1994) 'The Globalisation Challenge: Surviving at the Margins', *Third World Quarterly* 15:3, 427–445.

Molyneux, M. (1985) 'Mobilization without Emancipation? Women's Interests, State and Revolution in Nicaragua', in D. Slater (ed.) *New Social Movements and the State in Latin America* (Amsterdam, CEDLA (Interuniversity Centre for Latin American Research and Documentation), publication no. 29), 233–261.

Morales, E. (1989) *Cocaine: White Gold Rush in Peru* (Tucson, University of Arizona Press).

Morgenthau, H.J. (1967) *Politics among Nations* (New York, Knopf), fourth edition.

Morris, A. (1987) 'Regional Development, Trends and Policies', in D. Preston (ed.)

REFERENCES

Latin American Development: Geographical Perspectives (Harlow, Longman), first edition, 141–168.

Morris, A.S. (1981) *Latin America: Economic Development and Regional Differentiation* (London, Hutchinson).

—— (1995) *South America: A Changing Continent* (London, Hodder & Stoughton).

—— (1996) 'Geopolitics in South America', in D. Preston (ed.) *Latin American Development: Geographical Perspectives* (Harlow, Longman), second edition, 272–295.

Morris, A.S. and Lowder, S. (eds) (1992) *Decentralization in Latin America* (Westport, Praeger).

Moser, C. (1993) 'Adjustment from Below: Low-income Women, Time and the Triple Role in Guayaquil, Ecuador', in S. Radcliffe and S. Westwood (eds) *Viva: Women and Popular Protest in Latin America* (London, Routledge), 173–197.

Muir, R. (1975) *Modern Political Geography* (London, Macmillan)

Murillo Velarde, P. (1752) *Geographia de America* (University of Grenada), facsimile, Introduction by R. Ma. Sierra, 1990.

Murphy, C.N. and Tooze, R. (1991) 'Getting beyond the "Common Sense" of the IPE Orthodoxy', in C.N. Murphy and R. Tooze (eds) *The New International Political Economy* (Boulder, Lynne Rienner), 11–33.

Muscar Renasayag, E. (1989) 'Procesos de Urbanización y Modelos de Ocupación del Espacio en América del Sur', *Estudios Geográficos* 50:197, 621–659.

Nairn, T. (1977) *The Break-up of Britain* (London, New Left Books).

Nash, J. and Safa, H. (1980) *Sex and Class in Latin America* (New York, Bergin).

—— (1985) *Women and Change in Latin America* (South Hadley, Bergin and Garvey).

Navarro, M. (1989) 'The Personal is Political: Las Madres de Plaza de Mayo', in S. Eckstein (ed.) *Power and Populist Protest* (Berkeley, University of California Press), 241–259.

Navarro Andrade, U. (1957) *El Determinismo Geográfico de la Historia* (Quito, Talleres Gráficos de Ecuador).

Navarro Dávila, F. (1985) 'La Nación como Organismo Socio-humano Superior, frente al Estado, como Organización Jurídica y Política', *Revista Geográfica* (Instituto Geográfico Militar, Quito) 22, 65–73.

Newson, L. (1996) 'The Latin American Colonial Experience', in D. Preston (ed.) *Latin American Development: Geographical Perspectives* (Harlow, Longman), 11–41.

Nunley, R.E. (1970) 'The Role of Anglo-American Geographers in Latin American Development', *East Lakes Geographer* 6, 81–90.

Nunn, F.M. (1995) 'The South American Military and (Re)Democratization: Professional Thought and Self-Perception', *Journal of InterAmerican Studies and World Affairs* 37:2, 1–57.

Nye, J. (1990) 'The Changing Nature of World Power', *Political Science Quarterly* 105, 177–193.

O Tuathail, G. (1986) 'The Language and Nature of the "New Geopolitics". The Case of US–El Salvador Relations', *Political Geography Quarterly* 5:1, 73–85.

—— (1992) 'The Bush Administration and the "End" of the Cold War: A Critical Geopolitics of US Foreign Policy in 1989', *Geoforum* 23:4, 437–452.

—— (1993) 'The New East–West Conflict? Japan and the Bush Administration's "New World Order"', *Area* 25:2, 127–135.

—— (1994) 'Problematizing Geopolitics: Survey, Statesmanship and Strategy', *Transactions of the Institute of British Geographers* 19:3, 259–273.

O'Brien, P. (1975) 'A Critique of Latin American Theories of Dependency', in I.

REFERENCES

Oxaal, T. Barnett and D. Booth (eds) *Beyond the Sociology of Development: Economy and Society in Latin America and Africa* (London, Routledge & Kegan Paul), 7–28.

O'Brien, R. (1992) *Global Financial Integration: The End of Geography* (London, Pinter).

O'Donnell, G. (1973) *Modernization and Bureaucratic Authoritarianism: Studies in South American Politics* (Berkeley, University of California Press).

O'Donnell, G., Schmitter, P.C. and Whitehead, L. (1986) *Transition from Authoritarian Rule: Tentative Conclusions about Uncertain Democracies* (Baltimore, Johns Hopkins University Press).

O'Gorman, E. (1958) *La Invención de América* (México, Fondo de Cultura Económica).

Ohmae, K. (1990) *The Borderless World* (New York, Harper Business).

Ortega, E. (1983) 'La Experiencia Latinoamericana y el Desafío Campesino', in CEPAL, *Sobrevivencia Campesina en Ecosistemas de Altura* (Santiago, CEPAL), 21–69.

Ossenbrügge, J. (1983–85) 'El Desarrollo Reciente de la Geografía Política en el Area Anglo-americana', *Revista Geográfica de América Central* (Costa Rica), 21–22, 11–23.

Painter, J. (1995) *Politics, Geography and 'Political Geography'* (London, Edward Arnold).

Paravic Valdivia, S. (1988) 'Instituto de Investigaciones del Patrimonio, Territorial de Chile: Crónicas Antárticas', *Boletín Inpater* 1.

Parker, G. (1985) *Western Geopolitical Thought in the Twentieth Century* (London, Croom Helm).

Parker, R.G. (1991) 'Myths of Origin', in R.G. Parker (ed.) *Bodies, Pleasure and Passions: Sexual Culture in Contemporary Brazil* (Boston, Beacon Press), 7–29.

Parker, W.H. (1982) *Mackinder: Geography as an Aid to Statecraft* (Oxford, Oxford University Press).

Parsons, T. (1948) *The Structure of Social Action* (London, Routledge & Kegan Paul).

—— (1951) *The Social System* (New York, McGraw-Hill).

Paus, E. (1994) 'Economic Growth through Neoliberal Restructuring? Insights from the Chilean Experience', *Journal of Developing Areas* 29:1, 31–57.

Paz, O. (1985) *One Earth, Four or Five Worlds: Reflections on Contemporary History* (London, Harper Collins).

Pearce, J. (1986) *Promised Land: Peasant Rebellion in Chalatenango, El Salvador* (London, Latin America Bureau).

Peck, J. and Tickell, A. (1994) 'Jungle Law Breaks Out: Neoliberalism and Global–Local Disorder' *Area* 26:4, 317–326.

Pereira, A.W. (1993) 'Economic Development, Democracy and Civil Society: The North East Brazilian Case', *Third World Quarterly* 14:2, 365–381.

Petras, J. and Morley, M. (1992) *Latin America in the Time of Cholera: Electoral Politics, Market Economics and Permanent Crisis* (London and New York, Routledge).

Philip, G. (1985) *The Military in South American Politics* (Beckenham, Croom Helm).

Piel, J. (1989) 'Région et Nation en Amérique Latine: Le Cas du "Norte" Argentin (Tucuman, Salta, Jujuy) de 1778 à 1914', *Bulletin de l'Institut Français d'Etudes Andines* 18:2, 299–350.

Pinchas Geiger, P. (1970) 'The Development of Geography in Brazil', *East Lakes Geographer* 6, 56–63.

Pinochet, A. (1984) *Geopolítica* (Santiago, Editorial Andrés Bello).

REFERENCES

Pinochet de la Barra, O. (1987) *¿Puerto para Bolivia? Centenaria Negociación* (Santiago, Editorial Salesiana).

Pinto Serrano, D. (1986) 'Perfiles Históricos del Fenómeno Urbano Actual de Colombia', *Colombia Geográfica* 12:1, 7–21.

Poole, D. and Rénique, G. (1992) *Peru: Time of Fear* (London, Latin America Bureau).

Popke, E.J. (1994) 'Recasting Geopolitics: The Discursive Scripting of the International Monetary Fund', *Political Geography* 13:3, 255–269.

Potter, R. (1993) 'Little England and Little Geography', *Area* 25:3, 291–294.

Pratlong, G. (1982) 'Individualisme et Echange dans la Culture Andine Traditionnelle', *Bulletin de l'Institut Français d'Etudes Andines* 18:1, 23–53.

Pratt, M.L. (1992) *Imperial Eyes: Travel Writing and Transculturation* (London, Routledge).

Prebisch, R. (1965) 'Notes from the Standpoint of the Periphery', *CEPAL Review* 28, 195–206.

Preston, D. (1996) 'Themes in Contemporary Latin America', in D. Preston (ed.) *Latin American Development* (Harlow, Longman), 1–11.

Price, M. (1994) 'Ecopolitics and Environmental Nongovernmental Organisations in Latin America', *Geographical Review*, 84:1, 42–59.

Quijano, A. and Wallerstein, I. (1992) 'Americanicity as a Concept, or the Americas in the Modern World-system', *International Social Science Journal* 44:134, 549–557.

Radcliffe, S.A. (1993) 'Women's Place/El Lugar de Mujeres: Latin America and the Politics of Gender Identity' in M. Keith and S. Pile (eds) *Place and Politics of Identity* (London, Routledge), 102–117.

—— (1996a) 'Imaginative Geographies, Postcolonialism, and National Identities: Contemporary Discourses of the Nation in Ecuador', *Ecumene* 3:1, 23–43.

—— (1996b) 'Gendered Nations: Nostalgia, Development and Territory in Ecuador', *Gender, Place and Culture* 3:1, 5–21.

Radcliffe, S. and Westwood, S. (1993) 'Gender, Racism and the Politics of Identities in Latin America', in S. Radcliffe and S. Westwood (eds) *Viva: Women and Popular Protest in Latin America* (London, Routledge), 1–30.

—— (1996) *Re-making the Nation: Identity and Politics in Latin America* (London, Routledge).

Raffestin, C. (1980) *Pour une Géographie du Pouvoir* (Paris, Librairies Techniques).

Reboratti, C. (1982) 'Human Geography in Latin America', *Progress in Human Geography* 6:3, 397–407.

Reed, R. (1995) 'Household Ethnicity, Household Consumption: Commodities and the Guaraní', *Economic Development and Cultural Change* 44:1, 129–147.

Reid, M. (1984) *Peru: Paths to Poverty* (London, Latin America Bureau).

Reilly, C.A. (1995) 'Topocrats, Technocrats and NGOs', in C.A. Reilly (ed.) *New Paths to Democratic Development in Latin America* (Boulder, Lynne Rienner), 229–247.

Remmer, K. (1991) *Military Rule in Latin America* (Boulder, Westview Press).

Repetto, L.E. (1994) 'Women against Violence against Women', in G. Küppers (ed.) *Compañeras: Voices from the Latin American Women's Movement* (London, Latin America Bureau), 126–138.

Revista Geográfica de Chile 'Terra Australis' (1984–85) 'El Tratado de Paz y Amistad entre Chile y Argentina abre Caminos de Integración' (editorial), 28, 3–7.

Riesco Jaramillo, R. (1985a) 'Fundamentos de Geografía Política y sus Proyecciones Futuras en las Relaciones de Poder en la Cuenca Océano Pacífico', *Revista Chilena de Geopolítica* 2:1, 126–147.

—— (1985b) 'Fronteras y Tareas Geopolíticas Chilenas en el Océano Pacífico Sur y en el Continente Antártico', *Revista Chilena de Geopolítica* 2:1, 17–35.

—— (1985c) 'Geopolítica Austral y Antártica' *Boletín Antártico Chileno* 4:2, 14–17.

—— (1987) 'La Antártica: Algunas Consideraciones Geopolíticas', *Estudios Públicos* (Santiago) 25.

Riesco Jaramillo, R. and Santis Arenas, H. (1986) 'Las Fronteras Antárticas de Chile', *Cuadernos de Ciencia Política* (Santiago) 14.

Roberts, A. (1991) 'A New Age in International Relations', *International Affairs* 67:3, 509–527.

Roberts, B. (1978) *Cities of Peasants: The Political Economy of Urbanisation in the Third World* (London, Edward Arnold).

—— (1995) *The Making of Citizens: Cities of Peasants Revisited* (London, Edward Arnold).

Robinson, D.J. (1990) 'Liberty, Fragile Fraternity and Inequality in Early-republican Spanish America: Assessing the Impact of French Revolutionary Ideals', *Journal of Historical Geography* 16:1, 51–76.

Roccatagliata, J.A. (1988) 'Hacia una Geografía Prospectiva', in J.A. Roccatagliata (ed.) *La Argentina: Geografía General y los Marcos Regionales* (Buenos Aires, Editorial Planeta), 769–779.

Roddick, J. (1988) *The Dance of the Millions: Latin America and the Debt Crisis* (London, Latin America Bureau).

Roddick, J. and O'Brien, P. (1982) 'Europe and Latin America in the Eighties', in J. Pearce (ed.) *The European Challenge: Europe's New Role in Latin America* (London, Latin America Bureau), 15–40.

Rose, G. (1993) *Feminism and Geography* (Cambridge, Polity Press).

Rostow, W.W. (1960) *The Stages of Economic Growth: A Non-Communist Manifesto* (Cambridge, Cambridge University Press).

Rowe, W. and Schelling, V. (1991) *Memory and Modernity: Popular Culture in Latin America* (London, Verso).

Rowling, N. (1987) *Commodities: How the World Was Taken to Market* (London, Free Association Books).

Roxborough, I. (1979) *Theories of Underdevelopment* (London, Macmillan).

—— (1992) 'Neoliberalism in Latin America: Limits and Alternatives', *Third World Quarterly* 13:3, 421–441.

Rubin, J.W. (1994) '"COCEI in Juchitán": Grassroots Radicalism and Regional History', *Journal of Latin American Studies* 26:1, 109–137.

Rumley, D. (1991) 'The Political Organisation of Space: A Reformist Conception', *Australian Geographical Studies* 29, 329–336.

Russell-Wood, A. (1987) 'The Gold Cycle, c. 1690–1740', in L. Bethell (ed.) *Colonial Brazil* (Cambridge, Cambridge University Press), 190–244.

Safa, H. (1995) 'Economic Restructuring and Gender Subordination', *Latin American Perspectives* 22:2, 32–51.

Saint Lary, B. and Colantuono, M. (1993) 'La Patagonia en el Contexto de las Transformaciones Regionales', *Boletín Geográfico* (Universidad Nacional de Comahue) 19, 26–38.

Sánchez, J-E. (1992) *Geografía Política: Espacios y Sociedades 23* (Madrid, Editorial Síntesis).

Sandner, G. (1981–82) 'El Concepto Espacial y los Sistemas Funcionales en la Colonización Espontánea Costarricense', *Revista Geográfica de América Central* (Costa Rica) 15–16, 95–117.

Santis Arenas, H. (1985) 'La Geopolítica del Año 2000: El Poder Espacial', *Revista Chilena de Geopolítica* 2:1, 5–22.

REFERENCES

Santis Arenas, H. and Gangas Geisse, M. (1983) 'Los Primeros Pasos de la Geografía Política en la Cultura Occidental y en Chile', *Revista Geográfica de Chile 'Terra Australis'* 27, 9–32.

—— (1984–85) 'La Geografía de Chile en Cinco Obras Publicadas entre 1890 y 1962: Estudio y Análisis Crítico', *Revista Geográfica de Chile 'Terra Australis'* 28, 143–172.

Savard, M. (1992) 'La Catégorisation Précolumbienne de Cuzco, Perou', *Cahiers de Géographie du Quebec* 36:98, 299–318.

Schaedel, R. (1978) 'The City and the Origin of the State in America', in R. Schaedel, J.E. Hardoy and M. Scott Kinzer (eds) *Urbanization in the Americas from Its Beginnings to the Present* (The Hague, Mouton), 31–51.

Schattschneider, E.E. (1960) *The Semi-sovereign People* (Hinsdale, Ill., Dryden).

Scobie, J.R. (1972) 'Buenos Aires as a Commercial-bureaucratic City, 1880–1910: Characteristics of a City's Orientation', *American Historical Review* 77:2, 1035–1073.

Sempa, F.P. (1992) 'The Geopolitics of the Post-Cold War World', *Strategic Review*, Winter, 9–18.

Shanahan, E.W. (1963) *South America: An Economic and Regional Geography* (London, Methuen), eleventh edition.

Shanin, T. (1976) *Naturaleza y Lógica de la Economía Campesina* (Barcelona, Editorial Anagrama).

Shannon, T.R. (1989) *An Introduction to the World-System Perspective* (Boulder, Westview Press).

Shapiro, H. (1996) 'The Mechanics of Brazil's Auto Industry', *NACLA: Report on the Americas* 29:4, 28–34.

Sheahan, J. (1987) *Patterns of Development in Latin America: Poverty, Repression and Economic Strategy* (Princeton, Princeton University Press).

Short, J.R. (1993) *An Introduction to Political Geography* (London, Routledge).

Shumway, N. (1991) *The Invention of Argentina* (Berkeley, University of California Press).

Sidaway, J. (1992) 'In Other Worlds: On the Politics of Research by "First World" Geographers in the "Third World"', *Area* 24:4, 403–408.

—— (1993) 'The Decolonisation of Development Geography', *Area* 25:3, 299–300.

—— (1994) 'Political Geography in the Time of Cyberspaces: New Agendas?', *Geoforum* (special issue) 25:4, 487–503.

Sideri, S. (1993) 'Restructuring the Post-Cold War World Economy: Perspectives and a Prognosis', *Development and Change* 24, 7–27.

Skidmore, T. and Smith, P. (1989) *Modern Latin America* (Oxford, Oxford University Press), second edition.

Slater, D. (1985) 'Social Movements and a Recasting of the Political', in D. Slater (ed.) *New Social Movements and the State in Latin America* (Amsterdam, (Interuniversity Centre for Latin American Research and Documentation), publication no. 29), 1–27.

—— (1986) 'Socialism, Democracy and the Territorial Imperative: Elements for a Comparison of the Cuban and Nicaraguan Experience', *Antipode* 18, 155–185.

—— (1989a) 'Peripheral Capital and the Regional Problematic', in R. Peet and N. Thrift (eds) *New Models in Geography*, vol. II (London, Unwin Hyman).

—— (1989b) *Territory and State Power in Latin America: A Peruvian Case* (Basingstoke, Macmillan).

—— (1992) 'Theories of Development and Politics of the Post-modern', *Development and Change* 23, 283–319.

—— (1993) 'The Geopolitical Imagination and the Enframing of Development

Theory', *Transactions of the Institute of British Geographers* 18:4, 419–438.

—— (1994) 'Reimagining the Geopolitics of Development: Continuing the Dialogue', *Transactions of the Institute of British Geographers* 19:2, 233–238.

Smith, G. (1993) 'Ends, Geopolitics and Transitions', in R.J. Johnston (ed.) *The Challenge for Geography: Changing World, Changing Discipline* (Oxford, Basil Blackwell), 76–99.

—— (1994) 'Political Theory and Human Geography', in D. Gregory, R. Martin and G. Smith (eds) *Human Geography: Society, Space and Social Science* (London, Macmillan), 54–77.

Soler, R. (1980) *Idea y Cuestión Nacional Latinoamericana* (México, Siglo Veintiuno).

Somarribas, L. and Montero, C. (1983–84) 'Poder Municipal: El Caso de Costa Rica', *Revista Geográfica de América Central* (Costa Rica) 19–20, 89–116.

Springer, G.L. and Molina, J.L. (1995) 'The Mexican Financial Crisis: Genesis, Impact and Implications', *Journal of Inter-American Studies and World Affairs* 37:2, 57–83.

Spykman, N.J. (1944) *The Geography of the Peace* (New York, Harcourt, Brace).

St John, R.B. (1994) 'The Ecuador–Peru Dispute: A Reconsideration', in P.O. Girot (ed.) *The Americas: World Boundaries*, vol. 4 (London, Routledge), 113–133.

Stallings, B. and Horisaka, K. (1994) 'Japan and Latin America: New Patterns into the 1990s', in A.F. Lowenthal and G.F. Treverton (eds) *Latin America in a New World* (Boulder, Westview Press), 126–150.

Standing, G. (1989) 'Global Feminization through Flexible Labour', *World Development* 17:7, 1077–1096.

Stein, S. and Stein, B. (1970) *The Colonial Heritage of Latin America: Essays on Economic Dependence in Perspective* (New York, Oxford University Press).

Stepan, A. (1973) *Authoritarian Brazil: Options, Policies and Future* (New Haven and London, Yale University Press).

Stephen, L. (1992) 'Women in Mexico's Popular Movements: Survival Strategies against Ecological and Economic Impoverishment', *Latin American Perspectives* 19:1, 73–96.

Stöhr, W.B. (1975) *Regional Development: Experiences and Prospects in Latin America* (The Hague, Mouton).

—— (1981) 'Development from Below: The Bottom-Up and Periphery-Inward Development Paridigm', in W.B. Stöhr and D.R.F. Taylor (eds) *Development from Above or Below? The Dialectics of Regional Planning in Developing Countries* (New York, Wiley), 39–72.

—— (1987) 'El Desarrollo Económico Regional y la Crisis Económica Mundial', *Estudios Territoriales* 25, 15–25.

Strange, S. (1986) *Casino Capitalism* (Oxford, Basil Blackwell).

—— (1991) 'An Eclectic Approach', in C. Murphy and R. Tooze (eds) *The New International Political Economy* (Boulder, Lynne Rienner).

—— (1994a) *States and Markets* (London, Pinter), second edition.

—— (1994b) 'From Bretton Woods to the Casino Economy', in S. Corbridge, R. Martin and N. Thrift (eds) *Money, Power and Space* (Oxford, Blackwell), 49–63.

Strong, S. (1992) *Shining Path: The World's Deadliest Revolutionary Force* (London, Fontana).

Sullivan, L.E. (1985) 'Above, Below, or Far Away: Andean Cosmogony and Ethical Order', in R.W. Lovin and F.E. Reynolds (eds) *Cosmogony and Ethical Order: New Studies in Comparative Ethics* (Chicago, University of Chicago Press), 98–129.

REFERENCES

Sunkel, O. (1995) 'Uneven Globalization, Economic Reform and Democracy: A View from Latin America', in H-H. Holm and G. Sørensen (eds) *Whose World Order? Uneven Globalization and the End of the Cold War* (Boulder, Westview Press), 43–69.

Swyngedouw, E.A. (1992) 'The Mammon Quest: "Glocalisation", Interspatial Competition and Monetary Order: The Construction of New Spatial Scales', in M. Dunford and G. Kafkalas (eds) *Cities and Regions in the New Europe: The Global-local Interplay and Spatial Development Strategies* (London, Belhaven Press), 39–67.

Tanski, J.M. (1994) 'The Impact of Crisis Stabilization and Structural Adjustment on Women in Lima, Peru', *World Development* 22:11, 1627–1643.

Taylor, D.R.F. (1992) 'Development from Within and Survival in Rural Africa: A Synthesis of Theory and Practice', in D.R.F. Taylor and F. Mackenzie (eds) *Development from Within: Survival in Rural Africa* (London, Routledge), 214–258.

Taylor, P.J. (1993a) *Political Geography: World-economy, Nation-state and Locality* (Harlow, Longman), third edition.

—— (1993b) 'Geopolitical World Orders', in R.J. Taylor (ed.) *The Political Geography of the Twentieth Century* (London, Belhaven Press), 31–63.

—— (1993c) 'Full Circle, or New Meaning for the Global?', in R.J. Johnston (ed.) *The Challenge for Geography: Changing World, Changing Discipline* (Oxford, Basil Blackwell), 181–197.

—— (1994) 'From Heartland to Hegemony: Changing the World in Political Geography', *Geoforum* 25:4, 403–413.

Taylor-Dormond, M. (1991) 'The State and Poverty in Costa Rica', *CEPAL Review* 43, 131–149.

Taylor Pittman, H. (1986) 'La Geopolítica en los Paises del ABC: Una Comparación', *Revista Chilena de Geopolítica* 3:1, 53–104.

Tenorio Trillo, M. (1996) '1910 Mexico City: Space and Nation in the City of the Centenario', *Journal of Latin American Studies*, 28:1, 75–105.

Thomas, J.J. (1992) *Informal Economic Activity* (Hemel Hempstead, Harvester).

—— (1995) *Surviving in the City: The Urban Informal Sector in Latin America* (London, Pluto).

Thompson, A. (1992) 'Informal Empire? An Exploration in the History of Anglo-Argentine Relations, 1810–1914', *Journal of Latin American Studies* 24:2, 419–436.

Thrift, N. (1992) 'Muddling Through: World Orders and Globalisation', *Professional Geographer* 44:1, 3–7.

Thrift, N. and Leyshon, A. (1994) 'A Phantom State? The De-traditionalization of Money, the International Financial System and International Financial Centres', *Political Geography* 4, 299–328.

Timmons Roberts, J. (1995) 'Expansion of Television in Eastern Amazonia', *Geographical Review* 85:1, 41–50.

Tironi, E. (1985) 'A Reappraisal of the Role of Primary Exports in Latin America', in C. Abel and C.M. Lewis (eds) *Latin America: Economic Imperialism and the State: The Political Economy of the External Connection from Independence to the Present* (London, Athlone Press), 472–482.

Tironi, E. and Lagos, R. (1991) 'The Social Actors and Structural Adjustment', *CEPAL Review* 44, 35–51.

Todorov, T. (1984) *The Conquest of America: The Question of the Other* (New York, Harper & Row).

Toledo, X. and Zapater, E. (1991) *Geografía General y Regional de Chile* (Santiago, Editorial Universitaria).

REFERENCES

Torres, J. (1993) 'La Cooperación al Desarrollo y Efectos sobre el Territorio: Caso de la Cooperativa "Eddy Alonso" en Nicaragua', *Cuadernos de Ordenación del Territorio* 2:5, 35–44.

Torres Rivas, E. (1981) 'La Nación: Problemas Teóricos e Históricos', in N. Lechner (ed.) *Estado y Política en América Latina* (Mexico, Siglo Veintiuno), 87–133.

Townsend, J. (1987) 'Rural Change: Progress for Whom?' in D. Preston (ed.) *Latin American Development: Geographical Perspectives* (Harlow, Longman), first edition, 199–228.

Treece, D. (1990) 'Indigenous People in Brazilian Amazonia and the Expansion of the Economic Frontier', in D. Goodman and A. Hall (eds) *The Future of Amazonia: Destruction or Sustainable Development?* (London, Macmillan), 264–288.

Tuan, Y-F. (1977) *Space and Place: The Perspective of Experience* (London, Edward Arnold).

Tuyuc, R. (1994) 'From Grief Comes Strength: Indigenous Women's Resistance', in G. Küppers (ed.) *Compañeras: Voices from the Latin American Women's Movement* (London, Latin America Bureau), 111–116.

Ullman, R. (1994) 'The United States, Latin America, and the World after the Cold War', in A.F. Lowenthal and G.F. Treverton (eds) *Latin America in a New World* (Boulder, Westview Press), 13–27.

United Nations (1995) *World Economic and Social Survey, 1995: Current Trends and Policies in the World Economy* (Geneva, United Nations).

United Nations Development Programme (UNDP) (1995) *Human Development Report* (Oxford, Oxford University Press).

Urban, G. and Scherzer, J. (1991) *Nation-states and Indians in Latin America* (Austin, University of Texas Press).

Valle, N. (1983–84) 'Lineamientos Generales en Torno a la Formulación de un Marco Teórico y Consideraciones sobre el Estado', *Revista Geográfica de América Central* (Costa Rica) 19–20, 31–43.

van Klaveren, A. (1994) 'Europe and Latin America in the 1990s', in A.F. Lowenthal and G.F. Treverton (eds) *Latin America in a New World* (Boulder, Westview Press), 81–105.

Vega González, G. (1991) *Conversemos sobre el Geógrafo y el Ejercicio Liberal de la Profesión* (Santiago, Colegio de Geógrafos de Chile).

Vial Ruiz-Tagle, J. (1989) 'Ajuste e Interdependencia en América Latina', in Eyzaguirre, N. and Valdivia, M. (eds) *Políticas Macroeconómicas y Brecha Externa: América Latina en los Años Ochenta* (Santiago, CEPAL), 135–150.

Villarreal, C. (1986) 'La Supervivencia del Pueblo Awa – "Los Cabezones"', *Revista Geográfica* (Instituto Geográfico Militar, Quito) 23, 21–37.

Vitale, L. (1992) *Introducción a una Teoría de la Historia para América Latina* (Buenos Aires, Planeta).

Viteri Durand, J. (1994) 'La Geografía de la República del Ecuador', *Revista Geográfica* (Instituto Geográfico Militar, Quito) 33, 57–67.

Volbeda, S. (1986) 'Pioneer Towns in the Jungle: Urbanization at an Agricultural Colonization Frontier in the Brazilian Amazon', *Revista Geográfica* (Instituto Geográfico Militar, Quito) 104, 115–140.

Von Chrismar Escuti, J. (1986) 'Percepción de Algunos Problemas Geopolíticos de Chile: Su Trascendencia en la Seguridad Nacional', *Revista Chilena de Geopolítica* 2:2, 23–49.

Wachtel, N. (1971) *La Vision des Vaincus: Les Indiens du Pérou devant la Conquête Espagnole* (Paris, Gallimard).

Wade, P. (1986) 'Patterns of Race in Colombia', *Bulletin of Latin American Research* 5:2, 1–9.

REFERENCES

—— (1993) *Blackness and Race Mixture: The Dynamics of Racial Identity in Colombia* (Baltimore, Johns Hopkins University Press).

Wager, S.J. and Schulz, D.E. (1995) 'The Zapatista Revolt and Its Implications for Civil-Military Relations and the Future of Mexico', *Journal of InterAmerican Studies and World Affairs* 37:1, 1–43.

Wallerstein, I. (1974) *The Modern World System*, vol. I, *Capitalist Agriculture and the Origins of the European World Economy in the Sixteenth Century* (New York, Academic Press).

—— (1980) *The Modern World System*, vol. II, *Mercantilism and the Consolidation of the European World Economy, 1600–1750* (New York, Academic Press).

—— (1983) *Historical Capitalism* (London, Verso).

Waltz, K.N. (1993) 'The New World Order', *Millennium* 22:2, 187–195.

Ward, M. (1990) *The New Geopolitics* (London, Gordon & Breach).

Watters, R. (1970) 'Geography as a Social Science: A More Functional Geography for Latin America', *East Lakes Geographer* 6, 5–26.

Welch, J.H. (1993) 'The New Face of Latin America: Financial Flows, Markets and Institutions in the 1990s', *Journal of Latin American Studies* 25:1, 1–25.

West, R.C. (1979) *Carl Sauer's Fieldwork in Latin America* (Berkeley, UCLA School of Geography).

—— (1993) 'Aboriginal and Colonial Geography of Latin America', in B.W. Blouet and O.M. Blouet (1993) *Latin America and the Caribbean: A Systematic and Regional Survey* (New York, John Wiley), second edition, 34–81.

Whittlesey, D. (1944) *The Earth and the State* (Madison, USAF Institute).

Wilkins, M. (1970) *The Emergence of Multinational Enterprise: American Business Abroad from the Colonial Era to 1970* (Cambridge, Harvard University Press).

—— (1974) *The Maturing of Multinational Enterprise: American Business Abroad from the Colonial Era to 1970* (Cambridge, Harvard University Press).

Williamson, E. (1992) *The Penguin History of Latin America* (London, Penguin).

Wolf-Phillips, L. (1987) 'Why "Third World"? Origins, Definitions and Usage', *Third World Quarterly* 9:4, 1311–1328.

World Trade Organisation (1995) *International Trade: Trends and Statistics* (Geneva, WTO).

Wynia, G.W. (1990) *The Politics of Latin American Development* (Cambridge, Cambridge University Press).

Yagarí González (1994) 'The Right to Love and Politics: An Indigenous Activist's Perspective', in G. Küppers (ed.) *Compañeras: Voices from the Latin American Women's Movement* (London, Latin America Bureau), 138–146.

Yong Leng, L. (1982) *Southeast Asia: Essays in Political Geography* (Singapore, Singapore University Press).

Zaldívar, A. (1993) 'El Nuevo Orden Internacional y América Latina', in H. Casanueva (ed.) *América Latina: De la Marginalidad a la Inserción Internacional* (Santiago, Centro de Investigaciones y Promoción Iberoamerica-Europa), 9–14.

Zamora, R. (1995) 'Towards a Strategy of Resistance', *NACLA: Report on the Americas* 29:1, 6–10.

Zegers Santa Cruz, F. (1983) 'La Comunidad Internacional y la Antártica', *Boletín Antártico Chileno* 3:2, 3–7.

NAME INDEX

Abootalebi, A.R. 185
Agnew, J. 18, 77, 104, 106, 107, 184–5, 186
Agudelo Gómez, M. 51–2
Albert, B. 80, 81
Alvarez, E.H. 188
Alvarez, S. 155
Alvarez, S.F. 176
Amin, S. 18
Anderson, B. 171
Andreas, C. 124–5
Angotti, T. 146, 147
Aravena Ricardi, N. 65–6
Arnello Romo, M. 67, 68
Arocha Rodríguez, J. 162
Asis, E. 157
Astudillo Espinosa, C. 63

Baer, W. 25
Baerreson, D.W. 91
Baila, J. 160
Barham, B. 83
Barkin, D. 116
Barliza Zubiria, R. 137
Bassin, M. 17
Becker, B.K. 38, 135–6
Benito González, F. 139
Bergevin, J. 17
Berry, B. 15
Blake, G.H. 4
Blakemore, H. 20, 46, 58
Boateng, E.A. 3–4, 188
Boisier, S. 133
Bolívar Castillo, J. 70
Bowman, I. 4–5

Brandt Commission 3
Braudel, F. 5, 120
Brittin, A. 166–7
Bromley, R. 140
Bromley, R.D.F. 140
Brown, E. 24
Buchhofer, E. 99
Burdick, J. 162
Bustamente, F. 142–3
Butterworth, D. 147
Buzan, B. 75, 103

Camargo Pérez, G. 31–2
Campbell, B. 95, 116
Cano, W. 140, 153
Cardoso, F.H. 25, 184
Carnoy, M. 186
Carpenter, T.G. 61
Carrier, J. 84
Carvallo Ortiz, D. 31, 32, 160, 161
Castañeda, J.G. 75, 123
Castells, M. 147
Castillo, J. 70
Cavarozzi, M. 178
CEPAL 83, 90, 91, 96, 113, 148
Chamberlain, G. 60
Chance, J.K. 147
Chant, S. 112
Chilcote, R.H. 25
Child, J. 63, 142
Chisari, O. 84
Chomsky, N. 73, 118
Clark, M. 83
Clastres, P. 12
Claval, P. 20

229

SUBJECT INDEX

Printed in the United States
by Baker & Taylor Publisher Services